watchers at the pond

FRANKLIN RUSSELL

watchers at the pond

Drawings
by ROBERT W. ARNOLD

With a new introduction
by GERALD DURRELL

TIME Reading Program special edition
TIME INCORPORATED • NEW YORK

TIME
LIFE
BOOKS

EDITOR *Norman P. Ross*
EXECUTIVE EDITOR *Maitland A. Edey*
TEXT DIRECTOR *Jerry Korn*
ART DIRECTOR *Edward A. Hamilton*
CHIEF OF RESEARCH *Beatrice T. Dobie*

EDITOR, TIME READING PROGRAM *Max Gissen*
RESEARCHER *Joan Chambers*
DESIGNER *Ladislav Svatos*

PUBLISHER *Rhett Austell*
GENERAL MANAGER *Joseph C. Hazen Jr.*

TIME MAGAZINE
EDITOR *Roy Alexander*
MANAGING EDITOR *Otto Fuerbringer*
PUBLISHER *Bernhard M. Auer*

COVER DESIGN *Diane and Leo Dillon*

contents

FRANKLIN RUSSELL

editors' preface

The pond lies to the north, in that crowded borderland between zones where whippoorwills and warblers up from the South meet arctic owls, where northern larches share the woods with temperate maples. Here, in a place of forest and swamp, a meandering streamlet widens and the pond is formed. There are four watchers over it.

A red-tailed hawk looks down as he soars on wide wings that "soughed in the north wind, a sound that was rough-edged as he turned." From the rim of the pond, a snow-shoe hare stares warily, eying the enemy hawk overhead. In the water, an old muskrat, scarred by years of fighting, swims toward his shelter, on the lookout for marauding mink.

The fourth watcher is a man, and he sees beyond what hawk, hare and muskrat can see. He perceives the myriad, minuscule life that only men with microscopes and learning know about, and he populates his pond with hosts of strange little plants and animals, and some creatures that are both plant and animal. His pond is a lovely open place, yet crowded beyond counting—a place, he observes, where "a pinch of soil might enfold a million . . . animals."

Life goes on in curious ways around the pond. Tiny tunnels in the earth lead to hibernacular where earthworms, coiled around each other, sleep through the winter dry and safe—unless a mole pokes its way blindly into their dormitory. Ants jam into the corridors of their mounds, almost unmoving, turned by cold into an unthinkable contradiction: the slothful ant.

It is winter at the pond when Franklin Russell begins this book, which, in the years since it was published, has become a classic of its kind—a book touched at once with a poetic feeling for nature and a scientific understanding of its relationships, with an arithmetical delight in taking a census of the pond and an untiring appreciation of the plain and wonderful facts of nature. "One hundred and twenty thousand ladybirds piled together deep in one crevice," he reports matter-of-factly. "Several thousand were encased in clear ice . . . most would survive." He counts 7,000 mice of different kinds going about their business in underground quarters. He details the hygienic habits of chipmunks, who rouse themselves from hibernation to nibble sleepily at acorns in their storerooms and then, before going back to their sleeping rooms, visit a special chamber they have dug as a latrine. But Russell is also a dramatist who watches the creatures at the pond play out the endless bloody epic of survival.

His drama opens at a time so cold and harsh that even the crows have given up and flown southward. Chickadees and crossbills work frantically through the pines hunting for seeds. Snails suffocate under the thickening ice. A blizzard, tearing in, wrenches limbs from trees and exposes sleeping carpenter ants to mortal cold. The season shows no mercy and no sign of end—until at last there comes an east wind, miserably chilling but bearing rain. Water trickles through the snow and creeps under the ice. The sun suddenly asserts its warmth. Life at the pond stirs and takes on new pace. And so, now, does Franklin Russell's story.

That story came about in this way. Russell, a free-lance writer living in Canada, was called in by an editor of *MacLean's Magazine* in Toronto. "Russell," the editor said, "you know there's an awful lot going on in a single drop of water. Why don't you write about it?" Russell didn't like the idea (as a matter of fact, he admits that he never likes any suggestion from any editor). Tactfully delaying his refusal, he went away. He came back a few days later with his own idea. "Why just a single drop of water?" he asked. "Why not a whole pond?" And with the editor's acceptance, he want off to find a pond.

Without knowing it, Franklin Russell had been getting ready for this search and this story most of his life. Born in New Zealand in 1926 and brought up on a farm, he was altogether at home with nature. As a schoolboy he got excellent marks on his nature tests and papers, although he never studied particularly hard and never collected specimens. He grew up to dabble at many trades —truck driver, farmer, streetcar conductor—before he settled into newspaper work in New Zealand and

Australia, winding up in Canada and in the editor's office.

The drop of water that expanded into a pond produced a very satisfactory article, and then this book.

His pond is actually many ponds. Much of his observation was done at a park at Hamilton, Ontario. A dedicated supervisor had first set out to preserve a pond in its natural state and had then gone further, subtly adding to it so that it became a generality of all the ponds of the region. From there Russell went out to the Canadian countryside, to dozens of other ponds—to the lowlands laced with swampy creeks, to the deep forests of oak, beech and hemlock, and to the great plateau that stretched to the north.

Russell is not in the tradition of the older generation of nature writers—such as John Burroughs or Ernest Thompson Seton—who romanticized what they saw. He merges what he sees with what he has discovered in the library and laboratory. He can visualize single-celled organisms "by the billions in the pond . . . infinitely more varied than visible creatures . . . their soft unicellular bodies pulsing with slow and stately dignity." To such small creatures the amoeba—that familiar, form-changing protozoon of biology classes—seems a monster, gliding through the water, stretching out its pseudopod to capture and ingest them.

But the scientist does not get in the way of either the watcher or the writer. The ducks flying over the pond pass "very low and fast" and are "gone in the sound of a quack." Butterflies are making love "arched to meet in a strain of fulfillment." Again, a "loggerhead shrike swiftly impaled a wriggling leopard frog on the thorn of a pond-side bush. As the thorn passed through the frog, she screamed once, and instantly all the frogs hushed."

With spring, the cruel struggle to live engages all living, growing, passing creatures. The red-tailed hawk sees waterfowl flying by in the hundreds of thousands and, screaming at them from his high point in the sky, comes down to smash ducks greedily to earth in showers of feathers. Down the stream float sow bugs—80,000 in a single colony, by the author's estimate. Standing on their 14 legs on the surface, they reach the pond just as the crayfish burst out of their winter tunnels to devour them. A newly hatched dragonfly falls upon a tiger swallowtail butterfly, eviscerates it and, made incautious by its satisfied appetite, hovers too near a bullfrog whose long tongue flicks the dragonfly in. Mosquitoes hatch and start to feed—the males on plant juices, the females on the blood of animals and birds. Cattails grow fast, strangling other plants. Pond lilies spread their lovely leaves and keep the sun from reaching competing plants beneath them.

Over the battleground hangs the fragrance of anemone and wild strawberries. A male turtle patiently rides the back of a female, waiting until the exact moment when she is able and willing to be fertilized. Male mayflies, done with their delicate love-making, lose their wings and drop dying to the water.

With summer, the hunt grows more savage. Robber flies sting bumblebees to death. Mice squeal in terror as owls claw into their runways and nests. A hungry fox climbs a tree after a scared and bewildered squirrel—who gets away only to fall victim to the red-tailed hawk. The old muskrat roots at a mussel, lays it in the sun and, when the shell opens, eats the dying creature inside.

In the fullness of life, death is imminent. Russell is unsparing about the struggle—so unsparing that sometimes horror hangs over the idyllic pond. Here the author is in the tradition of the dean of all pond watchers. Thoreau recognized that the facts of life included death, and he reveled in this truth. Almost at the end of *Walden*, he wrote: "We can never have enough of Nature. We must be refreshed by the sight of inexhaustible vigor. . . . We are cheered when we observe the vulture feeding on the carrion . . . and deriving health and strength from the repast. . . . I love to see that Nature is so rife with life that myriads can be afforded to be sacrificed and suffered to prey on one another. . . . The impression made on a wise man is that of universal innocence."

The year at Russell's pond ends in a kind of innocence. The purpose of the struggle now seems clean and clear. Trees and plants spread their seeds. Crickets and katydids lay their eggs so that their species will winter over. The sky fills with southbound birds, then empties. One morning, the pond is stiff with a thin ice coating. The hawk wheels, takes a final look and flies off to watch elsewhere. The snowshoe hare, driven by the cold into a hollow tree, sits shivering, sniffing the damp air for danger. The grizzled muskrat, having finished his annual chore of dragging twigs to his nest, settles in and dies in his sleep of old age.

But in the quiescent cells that live on in the pond,

there remains a memory. It is, says Franklin Russell, "a haunting echo of the past . . . preserved . . . every detail of the spring awakening: the reaching for space, light, expression; the languorous heat of summer days; the slow waning metabolism of the last season; and the long sleep away from the sun. In every speck of living matter, there was this memory of the indestructible life force of earth."

—THE EDITORS OF TIME

RTP *introduction*

Gerald Durrell is a dedicated conservationist who lives on the island of Jersey in the English Channel, where he runs his own zoo, described in his book *Menagerie Manor*. His animal-collecting expeditions have resulted in a number of books, including *The Overloaded Ark, The Bafut Beagles* and *A Zoo in My Luggage*. The brother of novelist Lawrence Durrell, he has also written of their boyhood in *My Family and Other Animals*.

Nowadays it has become fashionable to believe that we are God and to forget that we are animals. I even saw an advertisement the other day for a learned work that promised to teach me all about the physical make-up of *Man and the Mammals*. I realize that I tend to get behind the times a bit but I felt that—even into my mental backwater—the news (for example) that all female members of the human race had suddenly lost the power to suckle their young would have penetrated.

Human beings are growing away from nature, to their

detriment. That is why it is refreshing to read a book like this, a book that shows the incredible and complex structure of even such a seemingly simple thing as a pond.

Those of you who still retain (in spite of television) the power to read, read this book and absorb its meaning. You will realize that a pond is not just a convenient place to throw your old tin cans or discarded bicycles. It is a highly complicated and beautiful web—but do not forget that it is a web of which you are a part. Multiply this story of a pond a millionfold and you realize that the world we live in is vast and intricate. Animals (yourself included) and plants are linked together like chains. Unfortunately, wherever you look, the weak link in the chain is man. Mainly rapacious, ignorant and greedy, man has inherited an enchanted garden in the world and is doing his best to despoil it. Like a willful child in a nursery, we are breaking all our toys. But toys can be replaced by the parents; are we sure that when we have finished wrecking the world God is going to buy us another one?

If this book does not show you the necessity for conservation in the world, the necessity for trying to understand the world better (for both selfish and esthetic reasons), then you are beyond redemption.

I once showed a very beautiful Brazilian toad to a very beautiful Englishwoman. The woman's charms I will not elaborate on except to say that, as well as being beautiful, she appeared to be intelligent and sensitive. The toad was a gorgeous patchwork of pink, silver and green, a design as wonderful as any carpet that came out of Persia. I felt sure that this intelligent woman would appreciate it as a work of art, quite apart from its obvious qualities as a toad. She peered at him for a moment and then said, "What use is it?"

By this she meant: What use was he to mankind? It is curious how an animal is judged by its usefulness and not its beauty.

"He's very decorative," I said, hoping she would redeem herself.

She sniffed.

"He also eats large quantities of insect pests and small rodents," I said.

"Ah," she exclaimed, looking at the toad almost affectionately, "then it *is* some good."

We will not let animals *be* for their own sakes. Ever since the publication of that bestseller of all times, the Bible, we insist on asking what use the animal is. If all the snakes in the world possessed the power to divine oil, they would be the most popular creatures on earth, but because they do not possess this power we revile them, forgetting that they are a most effective form of pest control. What use is an oil well when you are down with bubonic plague?

During the Great Plague of London, the authorities killed off all the dogs and cats, thus destroying, in a flash of brilliance that only comes to human beings, one of their best antidotes to the rats who were carrying the plague. This sort of ignorance of how the world functions is still common today, not only in primitive societies, where it is excusable, but in so-called civilized communities, where it is inexcusable.

This book will show you that the world you live in is a rich and wonderful place and it will also show you how little we know about it. If we are good gardeners and tend the world well, we will prosper. If we indulge in the bad husbandry we are indulging in now, the world will wither and fade—and we will fade with it. At the moment, our bad stewardship of the world is storing up for us something infinitely more frightening than an atomic war.

So read this book and learn. Learn that even a pond is important.

—GERALD DURRELL

watchers at the pond

the watchers in the snow

At the beginning of that year, a red-tailed hawk paused high over the frozen pond.

He wheeled slowly and the earth revolved under him. He saw the wing-shaped scar of the pond set in a still and fathomless variegation of bare trees, ravines, dark clumps of pines, and wandering white outlines of frozen streams. He floated under wide wings that soughed in the north wind, a sound that was rough-edged as he turned across wind, then smooth and keen as he cut down into its persistent force. His yellow legs bunched tightly under his belly, and his orange nostrils glowed in the cold sun. He knew a small part of the secret of the pond.

He saw the indistinct outline of a snowshoe hare standing at the edge of the pond. He heard the call of a dog fox and saw the bitch running through some trees. He saw the bare oak, north of the pond, where he had been born in a previous season. He saw a great marsh,

1

south of the pond, over which his sister and parents had headed to the ocean during the time of migration.

His extraordinary eyesight enabled him to see the pond in fine detail and to find movement in the stillness and meaning in flickering specks of color in thickets where tiny golden-crowned kinglets and chickadees darted between trees. He heard the tinkling, chippering cries of sparrows, rising like broken crystals. He saw the trunk circlers—woodpeckers, creepers, nuthatches—and heard their peeking calls and the precise ticks of their beaks driving into wood.

He saw a sparrow hawk speeding along the edge of the pond to a nameless destination. A male ruffed grouse burst out of deep snow with a stutter of powerful wings and drew a straight line of flight through the forest south of the pond. This mottled sweep of birches, beeches, and maples was silent and desolate. But the red-tailed hawk had seen flowers in its clearings and heard a score of bubbling, piping, whistling, chuckling cries of creatures living there in another season. He looked now and saw irradiant snow in a shaft of sun.

He wheeled over the marsh, which had only vestigial traces of the welling life it had held. Mink trails left shadows in the snow, fox prints crossed rabbit tracks, mice paths ran into snowbanks, and watching muskrats were black marks beside their shelters.

A black and white downy woodpecker flew over the pond and began drumming for borers in the hollow stems of horseweed stalks gripped in ice. The brittle

2

sound of his smashing beak sent a flock of redpolls scattering across the ice. An undulating charm of greenish-yellow goldfinches flew noisily over the pond and swerved down toward the angular wreckage of thistles standing up out of the snow.

The hawk turned slowly and flexed his great wings to maintain his height. In this cold wind, he flew merely to see and to travel. Gone was the exhilaration of fast-rising summer air carrying him so high into the sky's blue vacuum that the pond became a silver speck and the great southern lake dazzled him with a glaring slash of reflected sun. Gone also were the fall gales that could tumble him in massive turbulences of cold air meeting warm.

There was no urge now to scream with an excitement that would give pause to every ground creature hearing him. His hunting was so lean that all his senses were gripped by the need for food. He was trapped in a hostile winter, enduring it with strength acquired in another season, when blackbirds flashed red shoulders in the marsh, when ducks dropped low over the trees and smashed the sheen of the pond into dancing reflections, when rabbits and hares came down through the northern forest of oaks, hickories, and elms, passing from clearing to clearing and eventually disappearing into the dense overgrowth of water plantains, alders, sedges, and rushes that bordered the pond. The red-tailed hawk looked at the northern forest now but saw only a late sun coloring the snow.

II

The snowshoe hare knew another dimension of the pond. He waited by the ice, his breath fogging the air. He waited, watching the red-tailed hawk pivoting slowly overhead. He waited, smelling a fox. The hawk turned to the south, and the fox appeared from a thicket, plowing through chest-high snow and then disappearing into the shadows of the forest. The hare rose slowly from his crouch. Trembling, he sniffed.

The pond to him was an arena of dangers, an enclosure of leaning dark trees whence came ticks and whistles and scuffs of alarming sound. He trusted nothing. He had seen a pair of goshawks—great northern birds as large as he—floundering through the snow and brush on the tracks of cottontail rabbits and had heard screams when the rabbits were caught. Once, he had been watching a dog fox running across the far side of the pond when the bitch had leaped at him from the brush. He had hurled himself onto the pond, his bristled feet scrabbling for a grip on the slick ice as the bitch chopped her teeth close behind him. Being lighter than the fox, he had gained speed faster and shot up the northern slopes ahead of explosions of kicked snow.

The hare watched the woodpeckers drumming for borers and saw a shower of pine-cone petals falling from a tree top where a flock of crossbills were using their strangely shaped beaks to wrench seeds from cones. The hare could see, but not comprehend, other birds finding and eating spider sacs and moth cocoons that hung on twigs and bagworm cases that swung in

4

the wind. He watched the woodpeckers driving their beaks into tree trunks and pulling out sleeping grubs. He saw the mice sunning themselves at the entrances of their snow tunnels, which riddled parts of the snow around the pond. He saw an ermined weasel whisk into a snow bank and heard the terrified squeaks of

mice desperately scuttling through collapsing snow corridors as the weasel thrust after them.

He saw the dark, rigid shape of a dead crow standing on a branch in a nearby tree. On first seeing the crow in early winter, the hare had been wary, but later he had realized that the black shape was harmless. He could not know that the young and inexperienced bird had been caught in an exposed roosting place one night by a bitingly cold wind. The crow had wakened and looked over the moonlit pond through bleared eyes

5

and felt a paralyzing lassitude running through his body. He wanted to fly, but was held back by his fear of the dark. His powerful claws were locked on the branch in their sleeping grip, and he had died in the midmorning hours of darkness while slowly settling back on his perch. The morning sun flashed in his open eyes, and the hare bounded lightly away.

Like the hawk, the hare lived in a pungent aftermath of a beneficent season. South of the pond, he could see a great shoulder of granite with satellite rocks scattered around it, and in warm seasons he would lope among the rocks and cautiously crop succulent grasses. West of the pond, he could see a long ridge of snow rising from the ice. This was a rocky, sandy finger of land that jutted from the bank near where the inlet stream spilled into the pond. To reach patches of herbage on the peninsula, the hare might leap over the stream and graze, surrounded by the buzz and burr of thousands of crickets who lived in the sandy soil.

The hare, being a year-round resident, knew many of the ebbs and flows of pond life. He had seen thousands of creatures rising out of the water and spreading by wing and foot through forest and marsh. He had crouched in fear when forty thousand wings beat over the pond in a mass. He had seen hordes of dying creatures falling from the sky. He had heard singing mice and seen flying spiders. His senses were acutely tuned to the tempo of life at the pond because he was one of the hunted.

Looking over the petrified pond now, he saw a blunt,

grizzled head appear out of a small hole in the ice. The muskrat stood erect and sniffed the air. He saw the rabbit and ignored him. He also saw a weasel and slipped back into the water. The hare bounded silently through the northern trees, and the weasel was alone at the pond.

III

The muskrat swam under the ice toward his shelter near the sandy peninsula. A small fish fled into deep gloom. The muskrat was an old animal, graying at the muzzle, and his body was scarred by many escapes from hunters. A fight with a mink had left a long crescent-shaped scar on his right shoulder. Most of his right ear had gone, torn loose by an arctic owl three seasons before. He swam into shallowing water, past sticks and debris hanging from the ice, through odd effulgences of glowing light.

He swam over black slopes of mud and meadows of dark green grasses and past ribbons of old vegetation that soared to the ice roof and disappeared into it. He swam under a cluster of water beetles gathered around a bubble of air under the ice roof. He swam over semi-dormant dragonfly nymphs, the armor-clad aquatic youngsters of delicate flying creatures, and past a heap of leaves from which protruded two forked tails of fish that were drowsing the winter away.

For all his keen vision, the muskrat saw only an infinitesimal fraction of the microcosm of the pond. He saw worms so small they were whitish specks to his eyes, but he could not see the creatures the worms were pursuing and eating. His eyes caught only glimpses of the behemoths of this miniature world of the pond and missed the lilliputians, a million of which might not cover one of his eyes.

He swam on through this microscopic universe, unable to see the disaster that was decimating its creatures. The rotting vegetation and leaves on the bottom of the pond were exhausting the oxygen, and millions of creatures were suffocating. Even the fish were finding it hard to breathe, and some were dying.

The muskrat climbed a crawlway into the tumbled reeds of his shelter till he was above water level and inside a large dry chamber lined with leaves. He shook himself and lay down.

As he slept, perhaps he had a vision of easier times at the pond. The reeds would be thick and green, and the lambent air would be filled with floating seeds and flying insects. He would stand erect on a stone, watching dragonflies sweeping overhead, seeing the red-tailed hawk hovering high above them and hearing the movements of other muskrats among the reeds, some of them digging for shellfish, others pulling cattail roots from the mud. He would be fat, his pelt glistening sleekly in the sun, and his flattened tail would be instantly ready to hit the water as a signal of danger. At dusk he would watch for owls and for predatory mink, who roamed through the marsh. He would swim deeply through the pond and see sunfish and bass and dace moving among aquatic plants. He would see lizardlike salamanders and aquatic worms and snails and huge beetles. He would see hundreds of nymphs, the immature forms of insects who would later leave their drab aquatic bodies and fly above the pond. He would eat some of these nymphs, and as he crunched the tough exterior of their bodies, a kingfisher might smash into the calm water overhead in a shower of stars. The muskrat would see thousands of creatures heading out of the water as the sun streamed greenly down or as the moon halted at the surface. He would see a few of the millions of eggs that would be dropped into the pond. He would, in a long summer, see much of the pond's life. He slept now, remote from the fact and fantasy of abundant seasons, and the frozen pond lay silently around his rough shelter.

IV

In the middle of the winter equinox, the weather warmed briefly and the trees ran with water. Horseweeds faded spectrally into a rising mist, and a black squirrel kicked up a shower of snow as he raced to a tree. A pair of crows looked down somberly from an empty elm, and the red-tailed hawk dropped near a thicket and then flew by the marsh carrying a limp body. On the pond, a mink sniffed the muskrat's entrance in the ice, while the bitch fox watched her from the trees. A weasel glided over a hummock of snow and merged into the whiteness of the north bank.

As suddenly as it had started, the thaw was halted by a returning wave of arctic air. The warmer, lighter southern air was lifted under the wedge of the approaching cold, which rolled over the pond, dissipated the fog, and froze the trees. In the next flaming red dawn, the trees were shimmering and flashing like a phantasmagoria of prisms. The ice threw off multifarious hues—ambers, blues, pale reds—and these were broadcast so profusely by the red sun that for a day the pond seemed transformed from frozen immobility to glowing and sparkling motion.

But the winter marched on. That night, it destroyed the color. At dawn, the watery sun revealed motionless dark trees waiting for the spring.

The pond was the center of a universe that no one creature could comprehend. The red-tailed hawk knew some of it. He had hunted there in three seasons, but his knowledge was limited to one level of life at the pond. The old muskrat, who had survived six years there, knew another small part of it. A pair of arctic owls, who arrived at the pond as refugees from the really bitter weather of the far north, and brought with them an impression of white tundra, caribou herds, and the howling of wolves, had some special knowledge of it. The hunted hare knew fear in all the pond's seasons, but his life was only a fraction of this complex cosmos.

These creatures had neither the time nor the instinct to know all the incredible pond. The muskrat slept in his shelter. The rabbit crouched down in deep snow and the suffocation spread under the ice. The hawk turned the flat of his wings to the wind and sped south in a grand parabola and the pond dwindled into the distance.

the sleeping billions

The creatures suffocating in the pond could not escape; they were mute victims, and they were also a small minority compared with those who had found, through countless millennia of experience, that the safest way to survive the winter was to sleep.

A strange sleep it was. For some, it was a drowsiness always close to waking. The fish, buried among leaves and hiding between stones, made spasmodic efforts to hunt for food, but the bulk of the sleepers slept so abysmally they seemed dead.

Most of them were invisible. They blanketed mud and sand slopes at the bottom of the pond and clustered under the snow, sheltered in crevices, under bark, stones, leaves, and logs. They were swathed in mud and buried in the earth itself. They were present everywhere in stupefying numbers. A pinch of soil might enfold a million of the simplest animals on earth, the single-celled protozoans, and scores of thousands of the simplest plants, or algae, and a million parasitic plants, or funguses, and millions of bacteria.

Some of these sleepers were frozen solidly into ice or ice-crystalled soil. All had reduced their physical demands to a slow pulse or near to zero. Millions of them slept inertly as eggs, and there were eggs everywhere: stuck to submarine plants, concealed in sand, mud, and wood, laid under rocks and stones. These were tough, overwintering eggs laid by creatures who could not survive winter as adults. Some animals endured as invisible blobs of jelly. Others, equally small, were encased in brittle, hard shells, or loricae. At this level of life, the division between plant and animal life was often indistinct or contradictory. Some invisible plants would awaken to live like animals, moving purposefully and rapidly through the water, and some animals would live passive, plantlike lives.

Some of this multitude would crawl on simple feet, swim with primitive arms, or laboriously build portable shelters out of granules of sand. Some would look like thunderclouds, or like ducks, or like tiny sparkling suns. Some would live in colonies big enough to color the water in many brilliant hues; others would live independently. Some were so small it might take them years to move from the bottom to the surface of the pond. The movements of some were controlled by the sun. Many were indescribable. One sleeping animal, which even in clustered thousands was scarcely discernible, was intricately constructed of innumerable minute hexagons, each one capable of radiating prismatic colors.

All these creatures—never visible to a raccoon's eye—were complete, self-contained individuals. They would, in the seasons to come, move their brainless, sightless protoplasm in endless fighting, hunting, and eating.

This diminutive host now slept, merged into the muck of the pond bottom.

II

The pond was a dormitory, with creatures sleeping above and below the snow and ice. The rocks, sand, earth, swamp, rotting heaps of summer vegetation, deadwood, and mud of the pond were refuges for an incoherent agglomeration of sleepers. The ground surrounding the pond was tunneled by earthworms. Any creature who dug to deposit eggs or to bury food or build a nest broke through several of these corridors. They were numerous enough to aerate the soil, drain it, make it more friable and a richer, surer stimulus for the growth of trees and plants. Most of the tunnels ran horizontally at shallow depths, but some, dug in the late fall, plunged down deeply. Two, three, or four of these would converge into sleeping galleries, or hibernacula, and were filled with convoluted masses of sleeping earthworms. The solitary earthworm, who used only the shallow corridors he dug, needed protection from both the frost of the surface and the dryness of the subsoil. So he co-operated with his fellows to dig deeper tunnels and spent the winter in moisture-retaining colonies. The smaller burrows contained fifty to sixty earthworms; the larger ones, more than one thousand. About five hundred million earthworms were asleep round the pond.

Despite the depth of their burial and their immobility, the earthworms could not entirely escape the dangers of the season. A mole dug down the line of one worm tunnel under a tree and eventually broke into a hibernaculum containing three hundred and fifty sleeping worms.

For most sleepers, the biggest danger was freezing. Thousands of carpenter ants had insulated themselves from this by burrowing deeply into a dead pondside maple. For years, the ants had drilled corridors through the trunk which connected and interconnected, branched and rebranched till they made the interior of the wood porous. The ants slept bunched together in long, black motionless lines in these corridors, and piled into heaps in the galleries that punctuated the tunnels. They slept in a self-created microclimate, perfectly insulated from the freezing outside.

Other types of ants did not, or could not, escape the freezing. Instead, they had developed physical resistance to it, perhaps by antifreezing their body liquids, perhaps by partially dehydrating themselves before the freeze began. A snow-covered, bulky ant mound near the north bank of the pond was deeply frosted through its corridor-riddled earth. In places, earth and ants were frozen into large crystalline masses. Below the frost line, the mound was flooded with pond water and some ants were asleep in it. In the narrow space between frost and water, dense masses of ants were piled together, jamming corridors and galleries in glutinous black clots. Their metabolism was so low that their hearts were motionless, and they lived in a secret suspension puzzlingly remote from the hot activity of their waking lives.

The ants were deep sleepers. They shared this quality with the ladybirds, small orange-colored beetles with black-spotted wing covers who slept in faults and crevices that riddled the great chunk of rock on the pond's south bank. One hundred and twenty thousand ladybirds piled together deep in one crevice. They returned to this hibernating place every year, and many rock crevices were filled with the decayed remains of scores of thousands of them who had not survived previous hibernations. Like the ants, they could endure freezing and submersion. Some rain water had seeped into the biggest mass of them, and several thousand were encased in clear ice. They were still alive, and most would survive to the thaw.

III

But few other creatures could long survive freezing. During periods of prolonged cold, the frost line deepened and enveloped creatures with low frost resistance. In the shallows, the water was frozen to the mud and the ice had reached down and gripped some sleeping frogs. They might survive if the ice did not reach their hearts. Like all hibernating frogs, they sustained themselves by drawing in minute quantities of oxygen through the skin.

The frogs in the mud of deeper water were safe. Chorus frogs were buried among the tiny green spring peepers; cricket frogs lay with leopard frogs; spotted frogs lay around a lone bullfrog who had buried himself much deeper than the others. He dwarfed them all in size and strength and was the last survivor of a dozen bullfrogs who had once lived at the pond but who had been wiped out by a pair of herons nesting there one year.

The frogs would soon overflow the pond were it not for some other sleepers who had a considerable appetite for frogs. A hundred snakes hunted at the pond, though in thick grass, reeds, and water their numbers were never noticeable. Green snakes, racers, and rat snakes, and handsome red, yellow, and black king snakes slept around the pond. There were speckled water snakes and garter snakes, red-bellied snakes, some deadly copperheads, and constricting bull snakes. They would pervade the life of the pond. But now they slept, their black eyes closed, and their bodies coiled coldly in heaps of leaves in hollow trees, under logs, and in ground-hog burrows, in old bumblebee nests, and among crevices and slits in the granite rock.

Two thousand turtle eggs were buried in mud, sand, and earth. Scores of buried painted-turtle eggs waited for a touch of warmth to hatch in strips of sand at the edge of the pond. The adult turtles, each as big as a sparrow, had buried themselves in mud nearby. Map turtles with yellow-lined shells, some twice as big as the painteds, slept among showy box turtles and spotted turtles. The pond and its littoral held one hundred and fifty turtles, some as long as a robin's outstretched wings, others smaller than a hummingbird.

The salamanders, who looked like lizards but had froglike skins and were nocturnal amphibians, equaled the turtles in variety and excelled them in colors. They were hidden in every likely winter refuge. The biggest of them were the deep brown waterdogs, who had bushing red gills and were each as long as a duck's wing. There were eastern newts, with yellow bellies and red spots on their green backs; tiny red efts who were smaller than chickadees; and spotted salamanders who had solid, chunky bodies and bright yellow spots on their dusky black backs. The salamanders were shaded green, brown, yellow, and red—which, with assorted spots, were their base colors. They spent the winter in every possible stage of development. Some lived as slimy clots of eggs. Others lived as gilled larva and looked like fish except for their rudimentary legs. Some, in adult form, seemed near death, so deep was their hibernation. A few remained active under the ice.

IV

The survival of the cold-blooded animals was not as surprising as the endurance of more fragile creatures who, when awake, lived intensely and fleetingly. These lived and slept almost everywhere *except* in the pond.

18

The remains of a hornets' nest bulked gloomily over the snow in the upper branches of a dogwood bush near the sandy peninsula. It was a three-tiered complex covered by a gray, waterproof material. Inside were forty-eight dead hornets, still locked in the cells from which they had failed to escape in the late summer. Five thousand hornets had been born in this nest the previous year. The purplish-blue workers had whisked away across the pond, while the heavy-bodied drones hovered at the entrance of the ball-shaped nest. The workers had become drowsy and bedraggled in the fall and sought hiding places to die. All the hornets, except one, were now dead, and their bodies long since scattered into soil and water. The queen hornet, fertilized by a worker, had flown to the rock on the south bank to find a dry, deep crevice. Within her hunched, motionless body were six thousand separate lives of a future nest.

She shared the crevice with another sleeping queen, a large and rotund bumblebee, herself the only survivor of a nest of insects. She had been fertilized the previous fall also and would be among the first to leave this sleeping place and establish a new colony in an underground hive. By late summer, she might build it up to seven hundred bees.

A solitary overwintering queen was an efficient means for a colony to survive the winter, but it was not the only way. Several families of carpenter bees were asleep in the dead branch of an elm overhanging the pond.

They had returned to sleep in tunnels cut into the wood by their parents or by themselves during the previous summer. They had developed there as larvae. They lay one behind the other, the head of one bee almost touching the abdomen of the bee in front. Unlike the solitary queens across the pond, the females were not fertilized. One bee in ten was a male. This percentage was just enough to ensure fertilization of the females on waking.

Nearby, in dead goldenrod stems in the swamp, smaller, metallic-blue carpenter bees slept in similar tunnels cut through the soft pithy stalk centers. They slept head downward, in file and abreast, in places packed in so closely that the earliest wakers might be impeded in their way to the new season.

Also in the swamp were sleeping wasps, bumblebees, spiders, and beetles, hidden in old nests in the wreckage of the cattails. None felt the vibration of blows from questing beaks or heard the harsh clatter of dry stalks on windy days. They were insensible to the sibilant scratching of feet seeking a grip on stalks and to the soft throat noises of hungry birds communicating in the biting cold air outside.

No common equation of survival united the sleepers. They slept naked and clothed, buried and exposed, as adults and as eggs. Many flies, like bluebottles and greenbottles, unable to survive freezing, had pushed deeply into the porous insulation of rotten branches and logs. Others clustered in hollow trees. The flies were immensely varied and incredibly numerous; their larvae permeated water and leaves, and their eggs lay in bark and under stones. When the flies began emerging from their sleep, the pond would sound to the roar, rasp, whine, screech, drone, and rumble of their wings.

Most butterflies and moths slept in a cocoon or chrysalis stage of life in which they were changing, inside dark-colored tubelike cases, from caterpillars to butterflies, in a mysterious pupation process. Luna moths lay

in their cocoons in leaves and summer wreckage under the snow, and a field mouse, digging a snow tunnel, found one in some leaves and ate it. The luna caterpillars had wrapped themselves in leaves on the forest floor in the fall and were pupating there. Huge polyphemus moths, tightly cocooned in twisted leaves, lay scattered about the forest ground with the lunas.

Some pupae of butterflies slept in chrysalis cases that jutted from twigs and closely resembled the twigs in appearance. Great promethea moths slept as cocooned grubs in sagging, wrapped leaves in a grove of cherry trees. Red-admiral cocoons hung in ball-shaped houses, but for some reason as obscure as the hidden process that was so radically transforming this cocooned life, many red-admiral butterflies slept as adults in rotten logs and even under piles of leaves. Mourning cloak butterflies also slept as adults, their yellow-fringed wings spread in gloomy crevices of the rocks.

This intermediate stage of life survived the winter despite the fact that many chrysalises hung in full view of all the pond, suspended from twigs, branches, saplings, and hanging in piles of dried vegetation. They seemed to be the most vulnerable sleepers, and perhaps they were, for their mortality rate was high. Most mice, voles, woodpeckers, and chickadees hunted them eagerly.

A downy woodpecker came flatly across the pond into an old oak late one afternoon, saw a moth cocoon hanging at the far end of a fine-stemmed branch. He clutched clumsily down the slender stem and managed to peck the cocoon twice. But it swung around sharply on its retaining cord so that the blows had no effect on its tough cover. Eventually, the woodpecker flew away and the pupa inside the cocoon slept on.

The diverse, omnipresent beetles were other sleepers of boundless ingenuity, aggressiveness, and activity,

22

who, once they were awake, would dig, bore, fly, swim, be cannibalistic, herbivorous, and carnivorous. They had in common armored skins, six legs, and a range of brilliant colors. Painted hickory borer beetles slept in long tunnels in a grove of hickories. Innumerable locust borers, in larval form, slept in shallow galleries just under the bark of trees. The big sugar-maple borer beetles, handsomely black and yellow, slept behind the frozen sides of nearly every one of the five hundred sugar maples round the pond.

There were beetles hidden in woodpecker borings and buried in the sunken buds of pine trees, in hollow stalks and in rock cracks, and in disused carpenter-ant borings. Many of them had gone into the soil to escape the cold, and underground the May beetle larvae were fifty times their own length. Even at that depth, there was a hint of the quality of life that was soon to spread over the pond. Among the sleeping beetles were hibernating pupae of black digger wasps. These had reached the earth the previous summer as eggs laid by female digger wasps who had dug into the soil to find the beetle grubs. Their eggs had hatched into larvae, which had eaten many of the grubs, had pupated, and were now waiting to develop into winged creatures for the summer.

v

The obvious sleepers were the animals and insects. Less obvious were the sleeping seeds. There was no way

23

of estimating their numbers. At all levels of the pond's life—in the soil, water, and mud—seeds were waiting. Some had waited years to germinate and so were the longest, deepest sleepers of all. The hidden ground was absorbing a great many of them, including acorns, nuts, cones, single- and double-winged seeds of many types, seeds from berries and fruits. Many of them had special methods of warding off the teeming funguses and bacteria all around them and of awaiting a coincidence of events that would give them visible life.

A grove of elm trees the previous fall had liberated ten million winged seeds, which had rotated across the pond in chattering columns. They had fallen in water and on rock and had piled in drifts against banks. They represented the most profligate waste of pond life, because not more than a dozen of them would ever germinate. And the chances that these would grow into trees were remote. The pond was an ordered community of life, and the start of new life demanded first the destruction of the old. The seeds remained quiescent, but the trees would keep flying new crops of them over the pond in a ceaseless testing of the remotest chances to start new life.

All the trees were sleepers, even the evergreen white pines, spruces, larches, firs, hemlocks, and cedars, which kept their needled foliage throughout the year. Few other living things around the pond existed with such energy or with such insatiable demands on the resources around them. In their waking lives, they sucked incalculable amounts of water from the ground and extended roots and hair roots through the soil with such speed and power that subterranean creatures could sense the movement all around them. Cicada grubs would find hair roots entering their burrows, appearing out of the damp, glazed walls with almost perceptible speed. All this activity was now stopped, and the roots waited for the stimulus of spring to send them surging into a cohesive chain of movement to animate the highest branches above them. But now, the life force in them was collapsed to its lowest level.

VI

A colony of bats slept in the deepest crevice in the granite rocks, and their sleep was unique. In gray-tinted darkness, they hung in hundreds from the sloping roof, so closely packed together that the rock was concealed by their black, podlike forms. One bat was not breathing, and long moments passed before he took one shallow intake of air. He breathed ten times in an hour. Another bat was breathing once every two seconds. The breathing of a third bat began increasing rapidly. First she took one breath every second; then two breaths a second; then three, till her furred sides were pulsing in a blur. Her eyes opened and she stretched. Her activity seemed communicated to the surrounding bats, and their breathing quickened. The first bat silently dropped and flitted through the gloom, and others soon followed, till there were fifty or sixty of them darting up and

25

down the narrow crevice, their wings whispering softly but never touching as they passed and repassed. Then, as quickly as the flying had begun, it ended. The bats swooped up to the top of the cave, fastened themselves, licked their fur and wings briefly, and instantly relapsed into the drowsing of hibernation. The gloom of their hibernaculum contrasted oddly with the harsh light of the sinking sun outside, where the snowshoe hare crouched in the silent snow. Near him, a mass of ice crystals at the entrance of a hollow tree revealed the presence of other sleepers. The crystals grew from a continuous escape of warm air from two sleeping raccoons. The male lay on his back, curled up, with his eyes tightly covered by both paws. The female had jammed herself into a narrow space beside him and slept on her head, with her furry posterior pointing upwards.

Like the bats, the raccoons tended to be intermittent sleepers. They would soon rouse themselves and go into the snow to mate, and then resume their winter sleep. Their exceptional fatness, built up before they went into the winter sleep, sustained them, and their slowed-down bodies drew on this fat sparingly. A family of skunks soundly asleep in a burrow near the bank of the pond used the same sustaining device. They were seen occasionally venturing with distaste through the deep snow.

Some sleepers lived on stored food. The chipmunks, asleep in deep underground burrows, had dug storage chambers that they had packed full of acorns, nuts, seeds, and grass hay in the fall. They would rouse themselves occasionally, take a brief meal in the storage chambers, and visit a relief chamber where they emptied bladders and bowels before returning to the sleeping chamber to resume hibernation.

VII

All the sleeping creatures used mud, earth, wood, leaves, snow, ice, or water as insulation against the cold. But some creatures built their own insulation or carried it with them. Spiders endured as eggs or as dormant youngsters in egg cases, or as adults. Several thousand spiderlings and eggs waited under a rock face, perfectly hidden to outlast any weather. Some adult spiders had infiltrated piles of leaves and spun themselves into winter shelters of the finest and warmest silk.

The exceptional cold tested the insulation of all these materials during this winter. The deep snow and the bitter temperatures were gradually killing many creatures unprepared for such severity. The wind swept snow from one exposed shoulder of land on a northern plateau, and the earth, deprived of this cover, froze deeply. The ice was spreading through soil and leaf mould, and although some twenty thousand hibernating ladybirds there would survive this onslaught, beetles, flies, butterflies, and bees, sleeping contiguously, were dying.

It was a partial sleep for some. The old muskrat dozed in his pond shelter during the days. But down in the marsh, many muskrats had been made stuporous by the cold; the hunting mink swam under the ice, entered their shelters, dragged them into the snow, and killed them. The muskrats struggled feebly or not at all, their resistence to death broken by the cold.

For others, like the red-tailed hawk in the branches

of a bare oak, it was an uneasy and intermittent sleep. The hawk awoke and shivered violently, and these spasms of his muscles helped warm him a little. For the arctic owls, it was a warm sleep by day and night behind dense-feathered plumage.

The cold had caught several thousand aquatic snails in its solid clutch in the northern shallows of the pond. The snails could be seen clearly through the ice, fully withdrawn into their shells and sustaining themselves on tiny amounts of oxygen drawn through the ice. But this was a diminishing supply, and hundreds of them had already died.

The obscure dying of the snails was akin to another death by suffocation in the deeper waters of the pond. About one hundred fish—carp, sunfish, and dace—had died as oxygen in the pond diminished; their bodies rose out of leaves and mud and slowly ascended to the ice.

The numbers of the sleeping creatures were now evident. How could the pond cope with their awakening? What could prevent chaos as the sleeping billions stirred into life and began filling every part of the pond—its soil, mud, air, and water? What cosmic force would wring order and continuity out of a massive host of divergent life forces thrusting and straining in every direction? How could there always be a balance of these lives when their strengths and numbers varied so sharply? Why did not one creature—witness the layers of sleeping ants, or the clots of gallery-bound earthworms, the bulk of ladybirds—come to dominate the pond's life in a triumph of a single species? What was the purpose of such a multitude of species?

The sleepers, lying under the quiet snow, in gloomy water and constricting mud, posed many questions, and the answers to them all still did not reveal the secret of the pond.

the arctic storm

The winter moved to its final degree. The enduring cold was having a cumulative effect on the life of the pond so that here and there the resistance to death was weakening. The snowshoe hare paused near the pond, his face plastered with caked snow. He was shivering. He was having difficulty sustaining body heat with a diminishing supply of bark and buds. The red-tailed hawk ranged farther across country. He woke chilled in the night and saw an owl drift across the moon.

Two crows flew silently south one day, disappearing in mute recognition of the severity of the season. The cold was biting deeply into all life, quickening the consumption of winter food and fat and so narrowing the margin between life and death. It was an absolute time, the rigor of which might not be matched for another hundred seasons. The crows had sensed that there was worse to come as they had felt the pressure of the air

31

falling quickly in a preamble to an arctic storm. Soon after they had gone, the northern sky whitened, the overcast moved in, and the uneasy sound of rising wind was an audible dimension of new and bitter cold.

Many of the nonhibernating creatures were inhibited by this cold, but the chickadees and nuthatches, juncos and redpolls, grosbeaks and waxwings, cardinals and siskins had to keep moving. Their fast-beating hearts and high body temperatures would quickly fail without food. During the preliminaries of the storm, they hunted through tree tops and around tree trunks, dug in snow for seeds, cracked open pine cones, hammered into wood for hibernating beetles and sleeping grubs.

By midafternoon, it was so cold that the snowshoe hare was driven to shelter in deep snow among the dry chattering undergrowth near the sand and rock peninsula. The sky was now milky white, shot with streaks of black, and the wind groaned into the bare rattling forest. The first snowflakes fell as the overcast closed down tightly; the great arctic storm had begun.

As the snow, consisting of small, hard pellets, began streaming across the pond, all the overground creatures sought shelter. The birds headed for evergreens and dense thickets. The red-tailed hawk huddled in the racing gloom behind an oak trunk, and stray pellets ricochetted off his plumage. The wind whipped the fresh-dropped snow along the ground and piled it in long drifts. The pond, partially sheltered from the north by a tree-covered plateau and ridge, was a maelstrom of eddying snow. Above, the storm tore on to the marsh.

Shortly before dusk the snow changed in character. The pellets became mixed with large, damp snowflakes that stuck where they hit and turned to rime, forming on tree trunks, branches, stones, banks, and brush. The snow built massively into the force of the wind, crystal on crystal, and by dusk most of the trees were misshapen

with rime-ridden branches and trunks and their limbs crackled against the strength of the wind.

Even the arctic owls were restricted by the storm. The female clung to a branch in the lee of an oak. The male waited in a hollow tree south of the pond. He peered restlessly into the storm but could not see the mass of rime overhanging his refuge. It was slowly outweighing the strength of its anchorage, and when it fell, hitting

33

the ledge at the entrance, the owl was startled and jerked back sharply. For a moment he was rigid, his view of the storm obscured by a broken crystalline mass of rime blocking the opening.

A chunk of rime fell from a high branch and struck the leaf nest of a gray squirrel in the crotch of a hemlock tree. The squirrel shot from the nest and skittered up an iced branch, chattering in anger and doubt as he was enveloped in the writhing white columns of the blizzard.

A long-dead elm branch, which had been overhanging the pond for a hundred seasons, suddenly fell to the ice. The splintering of its base exposed a colony of hibernating carpenter ants. They were defenseless, their metabolism at its lowest point of the winter. Some were driven deep into the snow; others sailed free with the storm and whisked into the southern gloom. Their limbs twisted and froze in the cold. Some still clung together in attitudes of sleep, but unlike the mound-building ants, these creatures needed wooden protection against cold and had no chance of surviving the destruction of their shelter. Within moments the falling snow had erased all evidence of their existence.

The wind was forming long drifts wherever the ground caught its headlong, snow-laden passage for the briefest moment. At the beginning of the storm, a star-nosed mole had been impelled by hunger to leave his underground tunnel system near the swamp to go foraging along the edge of the pond. But on returning to the entrance of his tunnel into the earth, he found the territory made strange by drifts of snow. He quested back and forth, then plunged his starred snout into the snowdrift in an effort to break into the hidden tunnel. He dug into the drift, but the fine snow clogged his efforts. The pond darkened. The mole's search grew more urgent as he felt the increasing cold.

The storm reached into the evergreens for refugee birds, and they withstood it till a conifer branch, heavy with rime, swung down into some clinging juncos, which scattered into the hissing darkness. Once aflight, they had to keep flying, blundering through the fleeting bare outlines of trees and thickets. Some clutched and held to twigs, but the force of the storm pushed the rest of them south in a dubious quest for sanctuary.

The intrusive wind, in all its hostility, was shaped by the same cycle of birth and death that regulated the pond. At midnight it slackened, and almost immediately the thaw began. The rime fell and exploded in silent showers of white. The shivering gray squirrel, his leaf home ruined, had found temporary refuge in a hollow elm. The arctic owl had pushed through the rime and had flown off rapidly. The snow concealed the storm's numerous victims. When the dawn came at last, the rime was still falling into puffs of white smoke and odd branches cracked and fell. The pond had endured a long night.

II

At the zenith of the storm, winter seemed impregnable at the pond. In reality, it was about to disintegrate into an enveloping spring. The low temperature and the insistent north wind concealed the beginning of the new season, which was not arriving from space but growing from within the pond itself. Under thick ice in suffocating water, the dormant eggs of fairy shrimp had hatched into nauplii, tufted embryos each with a single eye. These moved to the undersurface of the ice. While the storm was still reverberating overhead, the nauplii hatched into twenty-two-legged creatures, which swam off on their backs through the dark water with rhythmic sweeps of upward-thrusting legs. They would grow quickly, assuming tints of green, pink, and bronze. Unlike other life, they would not be stimulated by warmth, and the thaw would kill them. For the moment they thrived in near-freezing water and would soon drop dormant eggs, which would turn into nauplii in the fall. As the pond warmed slightly, the fairy shrimps would sink, dying, to the bottom.

Elsewhere in the water, less contradictory life was preparing for the future. The six-legged nymphs of the May flies now sensed the approach of the time they would burst out of their aquatic bodies, their sex organs enlarged and their metabolism quickened, and fly from the pond. Many small nymphs and larvae hastened their growth in order to reach maximum strength and size when the thaw began.

Under snow southwest of the pond, a plant was forcing out of the iron-hard ground, its thick buckled leaves melting ice and snow as they rose. Two days after the end of the storm, it broke through the snow and emerged as a tip of dark on the white. It was a skunk cabbage, destined to grow wide-spreading green leaves.

Another hint of the illusiveness of winter had come toward the end of the blizzard, when a wild booming call had echoed repeatedly through the forest. Its authority and power chastened most who heard it. In the wake of the calls, the great horned owl that dominated the night forest might have been seen pumping silently through the diminishing snow, carrying the remains of a raccoon. His mate did not respond to his calls, perhaps because she was so incongruously visible, sitting tightly on a clutch of eggs in a roughly built nest in a large maple overlooking frozen swampland.

The owls, like the skunk cabbage, had moved into spring against the evidence of all seasonal signs. Cold, bitterness, arctic winds, sleet, snow, and ice were not related to the spring of the owls or the skunk cabbage. Both responded only to light. Since late in the previous year, the days had been lengthening, and when their light reached a certain duration, the owls automatically began their spring. They could ignore the weather and were quite independent of unseasonably early or late springs. The skunk cabbages were equally sensitive to the lengthening light. The plant under the thinnest layer of snow responded first; the plant buried deepest or hidden in the darkest shade would be the last to sprout. The latest skunk cabbages would emerge in the vanguard of the visible spring.

III

The arctic wind did not quite blow itself out. It dwindled, then veered and became freakish new wind from the west which shook the bare trees forcefully. During the darkest part of one night, anguished calls dropped from the black sky as creatures bound for far away destinations articulated their confusion. Their cries whipped away into nothing. At dawn the wind died, and its work was revealed. Hundreds of lost and exhausted birds were spread through forest and marsh. There were worm-eating warblers and hooded warblers, a western tanager, and nighthawks. They had been slowly moving north up the center of the continent, flying up the stream of a great river, when a violent gale swept in from the flat west. It tumbled them high in the air and sent them northeast at four and five times their normal flight speeds. Some dared to seek the earth but were smashed into trees or thrown into icy water. The survivors deposited at the pond had been wrenched from an instinctive pattern of behavior, and they reacted with piping lost cries and aimless flying. If they survived this violent change of life, they faced a migration beyond their instinct or experience. They had been heading for prairie country, but ahead now were rocky lands, thick forest, and deep lakes. Some of them might turn northwest in an effort to regain the original line of flight. They would face biting head-winds and wide stretches of open water. On the first day of the birds' arrival, two hooded warblers and the western tanager

38

died by the pond. Within a few days, the survivors had disappeared into the maw of the strange land.

IV

The omnipresent snow left no bold impression of its endlessly varied structure. But seen closely, it displayed minute and delicate patterns. Frost flowers had formed around the fringe of the pond as glistening, sparkling tufts of granular snow, and built up slowly, crystal on crystal, till they looked like miniature clustering white ferns.

Elsewhere, the wind had cut the snow into sharp-backed ridges, often with delicately ornamented sides or with sweeping figurations that bespoke its invisible force. In places on the pond itself, the wind had eroded the snow wilderness into three-dimensional relief, its layered structure showing clearly each facet of the snow's personality: the days of storm and silence, the periods of freeze and blizzard, the long and critical sleeping time.

Nowhere did the snow disclose its real structure, which was founded on one constant mathematical fact. Every particle of it was formed on some variation of six. Each was six-sided or six-pointed, six-armed, six-branched, or six-headed, and whatever the complexity of a flake each was a perfectly symmetrical unit. On some of the flakes each of the six points terminated in three points. Some of the flakes were double crystals, consisting of two flat filigrees joined by a center stem.

The snowflakes were symbolic of the hidden intricacy of the pond's physical existence in which the significance of six remained perpetually obscure, one of the many secrets of the pond. They fell, colliding in midair, smashing their delicate structure, then uniting on the ground into cohesive masses. In early winter the snow had been damp and heavy, and this snow lay close to the ground, holding in earth moisture. Mid-winter layers of dry and powdery snow, which had sparkled brilliantly in the sun, lay on top of the damp snow. Last was an amalgam of both damp and dry snow, with the damp predominating. Many creatures knew the amalgam contained a spring messenger, because the humidity of the new season had sweated into it and this could be measured by keen senses.

The snow layers held in the heat of the earth, making hibernation and shelter safer for many creatures. Before the storm, a partridge had come very fast across the pond, headed for a large snowbank, dropped suddenly, and dived into it. This looked like an accident, but he was safely buried under the snow, turning around in its fluffy substance and making an insulated sleeping chamber. The snow had fallen in behind, disguising his entrance.

The snowshoe hares dug shallow holes in the snow when the storm began and hunched out of sight as the snow piled over them. Cottontail rabbits crouched in the lee of fallen trees and rocks and let snow build over them till they disappeared. The deer mice, running through their mazes of snow corridors from one for-

40

aging place to another, were scarcely aware of the storm. They had their own nut- and seed-filled nests in hollow trees and in old birds' nests, anywhere they could more easily sustain body warmth. They kept constantly at work looking for more food and digging new snow tunnels, which they shared with red-backed mice and meadow mice and moles. The corridors, and a rich harvest of summer seeds and nuts, enabled more than seven thousand mice to survive at the pond.

The snow provided shelter and insulation, but it also provided hazards. The owls, the red-tailed hawk, the weasels, and sometimes the mink kept a mental record of any exit and entrance holes they saw in the snow. About twenty mice died every day while traveling from snow corridors to nests in bushes and trees. Panic would speed through the corridors when an ermine-coated weasel entered the maze and sought to corner a mouse in a blind corridor or trap him when the snow roof collapsed.

On some nights, a fox passing the pond would wrinkle his sharp nose at the smell of life under the snow. He smelled a sleeping partridge one night and padded to the scent with a barely audible scrunch of paw steps. But the sound wakened the partridge, and he tensed in his snow chamber. As the fox stretched delicately forward to where the scent was strongest, the bird erupted out of the snow and in one blurred moment fox and bird were together, the partridge hitting the side of the fox's jaw, then disappearing into the purple night ahead of the triple chomp of the fox's teeth.

The snow was a shelter, sometimes a treacherous one. The partridges occasionally misinterpreted the weather signs and would be sleeping under the snow when rain wet it and then froze it into a tough crust. The partridges would sleep till dawn, when some, more fortunate than others, would smash through the ice to freedom.

On one such night, the rain froze as it landed, fastening the wings of many juncos and chickadees to their backs. They tried flying at dawn but fell helplessly to the snow, where their lives depended on the remote chance of an immediate thaw.

The hunters used the snow with the same skill as did the hunted. On the morning the storm disappeared into southern murk, a cottontail rabbit dug to the surface through a drift that had deeply buried him in the night. He could scarcely tread down the snow to see above it. His whiskers were caked with snow. He tried to bound forward but sank back till only his ears showed.

Near the pond, within sight of the cottontail, the snowshoe hare was leaving his storm hideout. His bristled hind feet were more than twice the size of the cottontail's, and they partially supported his weight on the snow. The sun came out from behind a watery gray cloud, and the two animals were the only movements of life. The sky was clear of hawks, the branches free of owls. The cottontail plunged forward again.

Unseen by either animal was a watcher. He merged into a thicket, from where he could clearly see both of them, the cottontail on the hill among the trees and the snowshoe by the edge of the pond. The fox knew the

difference between the two animals. He knew his chances
of catching the snowshoe were remote; he lacked an
opportunity of surprise or ambush. Without hesitation,
he stepped out of the thicket and pushed across the

deep snow. Both creatures saw him instantly. The snowshoe froze for a second, then took one great catapult-
ing leap that drove his back legs deep into the snow and
sent him forward a short distance. His second great kick

increased his momentum till he was flying along the edge of the pond at high speed.

The fox ignored him and plowed up the hill through the trees toward the cottontail, whose struggles to escape his white imprisonment were futile. Through staring eyes he saw the fox's steaming breath and the snow clinging to his muzzle, and then he froze down and in that position was taken, killed, and eaten. From a bloodied circle, the fox flared his nostrils at a scent and barked sharply. The call went across the marsh, over the ridge into the north country, west up the frozen stream, east to the flatlands. There was no answer. After a moment, he struggled up the hill and disappeared toward the northern ridge.

The creatures in the pond steadily continued dying in the suffocating water. Above the ice, where only the keenest sense could detect the merging seasons, life was free to move to its greatest fruition. Shortly after dawn one morning, when the air was still crackling cold and the sky a milky white low overcast, a phoebe appeared on the southern bank of the pond. She was a small, greenish, upright bird, and in the gray light she bobbed her tail twice, then flew across the pond. She was working her way north to a nesting area she knew so precisely she would likely build a nest in the same tree she had used the previous year.

The moonless sky was broken by the harsh call of geese on ensuing nights. A sensitive ear could hear the rush of wings. Other birds appeared at the pond—swamp, vesper, and field sparrows—dun-colored birds who arrived unobtrusively and melted into the trees on the northern slopes. Their arrival was a part of the equinox.

The birds passed on, flying into light snow or the endless overcast. They were vigorous, hardy, enterprising, their instincts sending them farther and faster than those who would come after them. Their places were taken by new arrivals who followed in a straggling horde. Cowbirds appeared, walking jerkily along the margin of the ice, brown-headed and green-bodied, destined to leave their eggs in other birds' nests. There were more sparrows with red beaks and flashing white and black heads.

All the travelers suffered deeply, and some died. The sparrows left neat holes in the snow where they dived for food, seeking to anticipate the thaw and searching for winter crane flies which had pushed their wingless, spiderlike bodies deeply into the snow in search of places to lay their eggs.

A sparrow huddled in an ice-encrusted hickory tree one morning, her feathers puffed out. She stood on one leg and her eyes were passive and dull. She swayed, breathing stertorously. The flight north had been too much. She had failed to find sufficient food, her body heat was failing, and her will to live had gone. There was no chance of turning back. By midday, her head had disappeared into her fluffed-out back feathers. She fell from the twig shortly before dusk and left a round hole in the snow. In a dusky sky, birds called and disappeared north.

The pond was obeying signals from infinity, moving

in response to the gradual shifting of the earth on its axis as it tilted its northern hemisphere slowly toward the sun. It was imperceptibly obedient to vast aerial movements of water and air. Great masses of warm air were moving up from the equator and were being thrust back by the heavier, colder arctic air. But each time, the arctic air was less successful. The warm air would move above it, creating thick clouds and generating rain, which reached the pond as sleet.

The aerial revolutions of the mid-latitudes were well known to countless millions of migrant creatures long before the main effects of them were felt at the pond. The warm air kept rolling north over the continent. Hordes of migrants moved with and ahead of it, and the whole area was a turmoil of rain, sleet, snow, high winds, and flying birds. The beating wings, the wild sky calls, and the great northing movement communicated the urgency and expectancy of a new season.

An eastern rain wind began blowing tentatively across the pond. It seemed to have little authority, perhaps because it was so heavily laden with water vapor. But this vapor contained heat. As it swept over expanses of deep snow, only one of two things could happen. Either the east wind must be turned back by the cold or it must destroy the snow.

This matching of forces, silent and unseen, went on for half a day. The wind dropped its vapor everywhere into the snow, where almost instantly it condensed into water. It displaced cold from the snow, and the wind swept this away.

Finally, the snow was saturated with water. That night, parts of it began discoloring and collapsing. The wind's water ran down through it and over frozen ground to the pond. The bottom layer of snow disappeared. The mice hurriedly abandoned their snow tunnels and sought safety in trees as their corridors collapsed and disappeared. The thaw water ran out over the pond's ice, which was still concealed under snow.

That morning the sky cleared and the strangely bright sun sent the first palpable heat to the pond in one hundred days. The pond responded with a mist that grew from the cold succumbing to the warmth.

Beneath the ice, no great relief came from the suffocation crisis since the ice sealed the pond firmly against its own banks. But a small quantity of water seeped into the earth. It ran down worm holes, along roots, through shingle and shale. It appeared in the pond at disparate points as tiny bubbles, which began reviving a few of the suffocating creatures.

Above, the water was penetrating the pond's ice, washing it into long crystals, which met the fast-melting snow. Throughout the pond could now be heard the gurgle and skirl of melting snow and ice, the sound of running, flooding water.

Above all this, the east wind wavered. Its impetus was dying. The snowfields had taken its ocean water vapor, and its inland drive was gone. But the thaw had begun and would continue.

CHAPTER 4

the awakening

The pond had been asleep in ice, and it awoke in a flood. The brooks broke their frozen skins and the ice cracked and floated in the center of the pond. The reviving water poured in from all sides, and it stirred up turbid confusion which pushed the bodies of dead fish against the remaining ice cover and raced dead and living creatures downstream to the swamp.

Water ran down trees and flooded the top galleries and corridors of ant mounds and trickled into raccoon burrows and skunk holes. Gashes of black earth spread through the trees as the snow settled and retreated, revealing some of the winter's casualties.

The mole, who had never found the drift-hidden entrance to his burrow that night, had died of cold, with his head pressed against a tree root as though trying to move the immovable in his hopeless search. Bird and animal bodies near and far from him were a history of mortality in the area. The dead included a frozen cottontail rabbit coming into view in the remains of a thick snowdrift near the swamp. A weasel had killed her in midwinter and hidden the body in the snow. But the weasel herself had been killed by an arctic owl soon after, and now the thawing rabbit belonged to the pond's waking scavengers. Also appearing as sodden, shapeless masses of feathers were odd chickadees, juncos, a cedar waxwing with her crest still erected, and a lone Lapland longspur who had become separated from a flock heading south in search of food and had wandered disoriented until it paused to rest at the pond and there died of exhaustion and starvation.

Finally, a dull splotch of red emerged from one patch of snow that was sheltered from the sun by the granite rock. It was the drying, discolored breast feathers of an eastern cardinal, a last reminder of one creature who had briefly illuminated the pond's winter life before succumbing to it. Like the cedar waxwing, the cardinal had a head crest, which was also erected stiffly. In the cardinal's last moment of life, he had glimpsed a wheeling image of a sparrow hawk's glaring eyes and spread talons, and then he had been ripped off his perch and carried for a moment before dropping into the snow. The impact had driven him deep into a drift, and the sparrow hawk had flicked around in her own length to trace the prey before it disappeared. She hovered over the patch of disturbed snow, but she could not see the cardinal. After a moment's hesitation, she dropped into

49

the snow and floundered around trying to reach him. If she had paused, she might have heard him dying. His retching, convulsive breaths bubbled blood from each side of his reddened beak. But she gave up the hunt and flew away. Now, as the cardinal slowly emerged, the blood on his beak was black.

The survivors of the winter passed over the disintegrating cardinal in a thickening stream. No function of memory or instinct recalled his life. His mate of the previous summer alighted in a maple and looked down at the red blotch for a moment, then flew north across the pond. Her fellow voyagers cried out from the marsh and forest, and the nights throbbed with their whispering, whistling, shrieking, piping calls as they moved inexorably onward.

Actually or symbolically, all life was pushing out of its winter sheath for a place in the growing warmth. In the pond, a tiger salamander heaved himself out of the mud in a cloud of debris and swam sluggishly upwards. The sun fired the vivid yellow spots on his back while he was still deep in the water. He climbed onto a patch of ice fringing the pond, breaking through it several times before finding a footing. He was still struggling to dispel the sleepiness that had kept him under the mud for one hundred and fifty days, when he was seized and killed by a mink who seemed to know he was not fully awake. His death was a small incident. Following him was an avalanche of waking creatures.

Daily, the mud of the pond bulged and swirled as frogs kicked free and headed for their first lungful of spring air. Spring peepers, tiny frogs with piping voices, began singing. Skittering turtles sped through the water, looked at the sun with beady eyes, then swooped down in search of sleeping or waking larvae.

In the forest, a box turtle stirred in the earth under a tree. He pushed his head out of the crusty soil and jerked his legs feebly. As the ground cracked and fell away from his upthrusting head, he was temporarily blinded by the sun. Then he saw a skunk cabbage nearby, and in the distance wood ferns and bloodroots. He lurched slowly towards the pond through the burgeoning growth and stopped in a patch of saxifrage shoots. His shell was almost invisible in the multiple patterns of earth, plant, shade, and sun. He would wait there for other sleepy wakeners from hibernation—beetles, snails, spiders, flies—and would eat some of them as they passed.

The snails dropped in thousands out of the melting ice, some to lie and rot on the bottom, others to extend their viscid bodies and begin foraging. A bluebottle fly pushed his stubby body out of some wet leaves and groggily used his hairy legs to preen his wings. After an hour in the sun, he droned heavily away. From a leaf hibernaculum a centipede emerged. A mourning cloak butterfly jerked overhead, recently wakened from sleep in a hollow tree.

The blood of the pond was now moving everywhere. Hearts that had been beating intermittently began pulsing strongly. Protoplasm was pumping into action. The shadowy beginnings of new life were moving in

51

egg cases, seeds, pupae, larvae, burrows, caves, hollows, tunnels, in the buried tissue of perennial plants, and in the fiber of trees. The sodden earth sprouted a million shafts of grass. In a distant tree, a fox sparrow sang a tinkling song of exquisite and expectant feeling.

The trees reached down for strength to change the face of the pond. Their roots spread underground among the sleepers, and their branches expanded to meet the travelers. When the great horned owls began building their nest near the swamp and the skunk cabbages pushed up through the snow, the sleeping trees were rousing with them. Their lifelines, filled with motionless sap, tensed in response to the lengthening hours of daylight. In every root fiber, cells pulsed and thick, dark sap flowed upwards. The sap was drawn through roots and trunks and forced to the tips of every branch, pushing millions of buds from their protective sheaths. From root tips to topmost buds, capillary action created dynamic lines of force that were to transform the pond.

The rising sap and swelling buds burst into a panoply of tree flowers. The pondside alders dangled masses of greenish catkins into a cool south wind. A gold stain of catkins spread through the pussy willows. The black birches flowered, and their staminate catkins burst out of twigs in a slow, silent explosion. They lengthened rapidly and released pollen to fertilize the short green pistillate catkins. The tree pollen fell in showers, saturating the air in its blind search for the female organ. It mixed with other tree pollen, fell uselessly to the ground and the water, and floated to the horizon in prodigious abundance. The catkins hung singly and in clusters, slender and graceful, dense and profuse, fat and sensuous, a silent spectacle of a time of birth that permitted more sleepers to waken and join the pond life.

The awakening became a mass uplift of all life. A succession of warm days sent dogwood buds bursting

into newborn leaflets, and a hint of green blurred the stark outlines of deciduous elms. Hues of grass raced around the pond to cover the bare earth. Plants pushed up from under the debris of the forest floor. The first green tips of toothworts, partridgeberries, bunchberries, and foxgloves appeared. The broad green leaves of the trilliums seemed to leap into sight, followed by the rue anemones and bugbanes, and handsome bellworts with dark green leaves. The full, nodding leaves of yellow violets shortly preceded their delicate five-petaled flowers.

The profusion of plants filled the eye as they drove out of the earth in an astonishing variegation—orchids and adder's-tongues, Mayflowers and bedstraws, saxifrages and pipsissewas. The precursory skunk cabbages had now spread enormous leaves and were flowering. The blossoms were packed with drinking gnats and sluggish flies and bees. The quick-growing bloodroots also had early flowers, delicate bluish white blossoms, each of which appeared on a slender stem rising from the enclosing embrace of a single green leaf.

II

A host of creatures were rousing in the crevices of the granite rock. A queen bumblebee stretched her legs and wings. She had been wedged against two narrowing faces of rock deep in the crevice, and her body fur was disheveled and ragged. With her metabolism quickening, she breathed deeply and stretched again. Then she droned waveringly toward the bright light at the entrance of the crevice. Though weak, she unhesitatingly launched herself into the pastel lighting of the pond, flying low over the water and feeling the chill of a steady easterly wind. On the far side of the pond she saw the tree catkins and headed up toward them. She thumped heavily onto the first catkin, sending it swinging wildly and showering pollen, bumped to the next flower, and when she found no nectar in the third catkin she shook it in seeming anger and buzzed loudly. Within an hour, she was smeared with pollen and her nectar-hunting had become an orgy of eating. Then she headed along the bank of the pond, flying low, buzzing into piles of leaves, swinging around tree trunks, hunting out every crack and crevice in the ground, putting her head into mouse burrows, crawling hastily into hollow roots. She found a shrew's burrow and was crawling into the entrance when the shrew rushed from below and bared his teeth at the bumblebee. The queen shot backwards, and the shrew watched her fly out of sight.

While the bumblebee continued her search for a burrow in which to begin a new colony, other queens were waking. From the crevice that had sheltered the bumblebee came a lean-bodied wasp, who flicked rapidly over the pond, examining the wreckage of old nests in bushes and trees. Other wasps and hornets appeared to begin a purposive examination of the pond.

In a hollow tree filled with rotten wood, thirty-eight thousand drowsy, fertilized female mosquitoes began stirring. They would waken slowly, but when they emerged, they would urgently search for their first meal of blood, obtained perhaps among the feathers of a sleeping chickadee or hawk or owl. This would prepare them for the task of laying floating clusters of eggs in still patches of water around the pond.

Winter had created underground chaos in the ant

mounds, and the ants were already laboring to clear jammed and collapsed tunnels and galleries, working closely behind the retreating ice. In the lower levels of one big mound near the pond, ants were struggling awake underwater, clawing up through flooded tunnels till they reached dryness. They joined in the work immediately without attempting to go aboveground. Newly excavated earth granules spilled out of mound entrances. One ant mound lay under a rivulet of thaw water from higher ground. The ants struggled to stem it, but the water soaked into the absorbent mound, carrying ants with it and driving many of them deep into the soil. The earth tunnels and galleries collapsed. The heart of the nest was destroyed, and only stray ants were left, wandering homeless and aimless on the surface.

The underground struggle to awake was most urgent for some earthworms. In their deep hibernacula, most of them were safe from the flooding thaw because the shallow complex of tunnels dug the previous year drained most of the upper ground. But some sleeping galleries were flooded. In the face of the incoming water, the worms roused and struggled against it to reach the surface. They could withstand some immersion, but eventually water would drown them. They sprawled in great numbers aboveground, and migratory birds harvested them.

Life stirred vastly in the marsh and swamp. With the ice all gone, early ducks dove for small aquatic plants that rose from the muddy bottom. A tinge of green

colored the banks and promontories as reeds sprouted and grass shot from the black ground. The snowshoe hare raced with the new growth. He was changing color, too, turning blotchily from white to gray and brown. Through the new growth strode coots, dark, ducklike birds with unwebbed feet and white bills, and other water birds such as gallinules, bitterns, and herons. Two kingfishers appeared one morning and dived for fish.

Soon all the marsh was suffused with green. It was a green of marsh calla, blended with pink persicaria and the broad flat leaves of reeds. The marsh was filling with the noise of chackering red-winged blackbirds. Marsh wrens swayed and chattered on cattail stalks.

Swamp sparrows trilled. The walking ground birds—virginia and sora rails and others—skulked through the low masses of new growth. A sandpiper trilled with a delicate, slender sound. Then he began to sing in a haunting, ethereal whisper. The old muskrat looked young and sleek, wrinkling his grizzled snout at the smells of the pond while standing erect on his rough shelter. Perhaps his eyes were caught by new colors at the pond. The warblers were arriving and setting their colors against the fresh green of new leaves. Yellow warblers moved brilliantly in the sun. The blues, greens, grays, and oranges of a multitude of other warblers—myrtles, redstarts, bay-breasteds, water thrushes, oven-birds—encircled the pond.

The earth-bound eye could see only part of the arrival of the birds. At midnight a nighthawk, who had left a southern continent and crossed a sea to reach the pond, flew with other migrants across the great lake. He heard the air around him filled with the whisper and rustle of wings. In deep darkness he heard calls blended in grada-tions of urgency, fear, and inquiry. The calls wheed, mewled, wheated, twickered, keeped, kewed, purred and chattered. The nighthawk saw a moon reflection and curved down toward the pond. Above him streamed countless birds who would never know the pond.

During these days, the pond sheltered thousands of night migrants. They settled into trees at dawn, their numbers and colors giving a brilliant indication of the diversity of life spreading into the north. Their colors and sounds soon became a part of the pond, which glowed ceaselessly with the golden tints of yellow-breasted chats, the brilliant reds of tanagers and rose-breasted grosbeaks, the varied colors of golden-winged and blue-winged warblers, cowbirds, thrushes, owls, sparrows. There seemed no end to the travelers.

But the migration had only begun. In pink evenings, the sky was marked with flowing lines of ducks and geese moving north. On some days, the marsh held one hundred thousand birds, who took to the air in sinuous columns and clusters and clots. Their powerful wing beats created rushing sighs of movement over the pond. Some ducks passed over very low and fast and were gone in the sound of a quack. They left hints of their passage: spiraling feathers colored deep purple and shot with touches of red from the straining wings of mallard drakes, or perhaps golden- and green-hued and red feathers from wood ducks. The ducks rode across the rays of setting suns into the north.

Less obvious were the crows. They dribbled north singly and in groups, and they came continuously, fed from southern flocks numbered in millions. The red-tailed hawk watched the fliers from a high point in the sky and screamed at them. He smashed ducks to earth in showers of feathers. He saw red-winged blackbirds materializing out of the nights and scores of thousands of them settling in the marsh and swamp. Their evening flights decorated the ground beneath him. One evening a flock of them came hurtling out of the red sky and settled with a rush of wings. The pond became uproarious, a cacaphonous, squabbling mass of red-shouldered birds, when, without warning, they roared aloft again and were gone in a disciplined mass, headed into the sunset and unknown horizons.

III

The awakening around the pond was spectacular. But its effect diminished when compared with the awakening within the pond. The stream feeding the pond brought many sow bugs or isopods, each as big as a large fly, with fourteen legs sprouting all along the length of a thick-stemmed body. They sank toward rotting vegetation, on which they would feed. There were scores, then hundreds, then thousands of them. An upstream colony had been broken up by the thaw, and eighty thousand sow bugs came into the pond in one day. Their arrival coincided with the emergence of crayfish from long tunnels dug into mud and earth around the pond. The crayfish burst through earth plugs they had built in the fall, and about two thousand of them began eating the isopods. Great diving beetles rose from the mud around them and went gliding effortlessly through the water with sheeny purple skins reflecting the spring sun. From out of slimy vegetation came flat, bronze-colored whirligig beetles, kicking swiftly through the water with long hairy back legs. They headed for the surface and jerked back and forth there. One of their eyes watched above water, the other watched beneath the surface.

The air-breathing back swimmers, who had clustered thickly in the vegetation of the pond during the winter in search of traces of oxygen, swept upwards and sent their tapered, cylindrical bodies shooting across the surface with thrusts of their long back legs. The water boatmen, blackish creatures also with sweeping propulsive legs, emerged from rubbish around the periphery of the pond, where they had barely survived on traces of oxygen trapped there in the freeze. With them came water striders, slender-bodied creatures with six long hair-tufted legs. They jerked to the surface, burst through the water skin and then, standing in six dimples, darted across the pond. Other water striders emerged from leaves in the forest and flew clumsily to the water.

The bottom of the pond was now alive with the crawling nymphs of creatures who would later leave the water on wings. The powerful six-legged dragonfly nymphs lurked in ambushes, many of them waiting for damsel fly and May fly nymphs, who were dribbling into the pond from the inlet stream.

The appearance of great numbers of nymphs, beetles, and bugs coincided with the multiplication of invisible and incredible aquatic animal and plant life around them. Thousands of tiny wolffia plants, which had sunk to the bottom to overwinter as almost invisible spheres, slowly floated to the surface. They had consumed their winter starch food and now rose to mass together in an almost impenetrable community of tiny green leaves.

The diversity of life in this miniature universe seemed infinite. The statoblasts, or seeds, of some peculiar animals, the bryozoans, fastened disc-shaped bodies ringed with hooks to almost any object they touched. They

61

grew, splitting their two shell-like discs to allow a pulpy body to grow between, one end of it sprouting a waving mass of cilia, or hairs. These individual sprouts of life would, as the spring advanced, form tubelike corridors fastened to leaves, stones, or sticks. The tubes would join each other, forming colonies of bryozoans, till eventually they became thick and visible in the pond, looking like many mantles of green.

From soil and mud, from under stones and around roots, came the bristleworms, their tubular bodies sprouting hairs sparsely or in clusters, depending on their type. They headed for the wreckage of decaying plants and would spend the rest of the unfrozen year feeding on them. Their numbers were great but were exceeded by the host of pear-shaped crustaceans, the copepods, hatching from invisible nauplii, the intermediate stage of their development between egg and adult. They were turning into blunt-nosed swimmers with trailing tails, faintly resembling in miniature the crayfish and lobsters to whom they were related. Some had wide-ranging feelers, some had no legs, others had eight. The copepods grew as the ostracods, also crustaceans, began rising from the mud. They were like tiny shellfish, with two muscled shells that could send them gliding speedily and gracefully through the water.

Throughout the fast-warming shallows, a swarm of miniature plants rose from debris at the bottom. Desmids headed for the warmth of the new sun and fastened themselves to the stems of plants. Collectively they looked like a fine green film. Their soft, unicellular bodies pulsed with slow and stately dignity. They existed in an indescribable profusion of shapes and sizes, some of them having smooth spherical bodies, others being finely striated or roughened with dots and points or long slender spines. They sprang to life in billions.

Among them were some diatoms. Their greatest uprising was in the swamp waters to the west, but their emergence everywhere was spectacular. Unlike the unicellular jelly blobs of the desmids, the diatoms had hard exterior cases, which, when eaten by beetles or nymphs, would crack and the yellowish-brown contents would spurt out. They moved rapidly, changing direction sharply, endlessly reaching for space and light. Despite their billions, few pond creatures could see them in the mass, much less discern their body decorations, which consisted of innumerably various dots and circles, hexagons, and transverse and longitudinal lines. If seen closely in the sun, their brownish bodies, filled with a slow-pulsing protoplasm, refracted the light softly, showing liquid moving in mysterious perpetuation of simple life.

The rise of these organisms was vital to the survival of many of the pond's creatures. They were a prime source of life, a host that would feed a multitude. Together with many other equally small creatures, or plants, they would form the algae of the pond, which would appear later as slime on rocks and brown stains on plant stems as their numbers multiplied again and again.

In the deepest part of the pond, the bullfrog rose mountainously out of the mud and kicked sluggishly to the surface. He swept unseeing past a group of tiny and delicate infusoria, which were animals but behaved like plants and which had barely survived the oxygen crisis even though dormant on the bed of the pond. During the suffocation most of them had disintegrated, but now the survivors were poised to re-establish themselves in the pond. They lived in all the expected variety of the microcosmic world, some protected in shells, some unprotected as naked individuals, and some collected into communities that looked like tiny branching trees. When the bullfrog passed some of these, they yanked down their branches faster than the eye could follow. When the bullfrog had gone, the creatures uncoiled from protective balls and resumed feeding.

At the level of life of the infusoria, the modes of existence were echoes of primeval life going back hundreds of millions of years. A volvox came spinning through the water, a tiny sphere of jelly propelling itself with waving cilia sprouting from its body. Each cilium belonged to a separate single-celled creature; yet in some incomprehensible way these creatures had united into a community organism bound together by protoplasmic threads.

The rhizopods, the animals lowest on the scale of pond life, or life anywhere, were beginning their race for space. They were rising like ghosts from the bottom, ten thousand billion blobs of jelly that had been asleep and invisible. They proliferated in the warming water by division and redivision of their single cells. The amoebae were the commonest rhizopods, but they had many relatives and their rise coincided with the appearance of rotifers, microscopic animals whose transparent stem-like bodies revealed their simple, coiled guts. They were emerging in clouds from dormant winter eggs. A few of them were just visible. One came through the clear water propelled by what appeared to be two small rapidly rotating wheels mounted on his head. The wheels were two discs of vibrating hairs which created such strong currents that the rotifer could swim and catch prey with them.

The winter suffocation had scarcely harmed the entomastraca, and now they began to teem. They were miniature crustaceans, and some were visible as innumerable skipping, jerking, floating white specks. Their greatest time was yet to come. They were the scavengers and purifiers of the pond, and they would transform poisons into food for many creatures.

Water fleas appeared with water mites and many different worms. Planarians, black flatworms who looked like leeches, undulated gracefully through the water. Scores of thousands of tiny tubifex worms protruded waving, questing heads from projecting towers, studding some parts of the bottom of the pond.

All this life—visible and invisible—appeared against countless blunt shoots springing from the bottom, from mud, from old roots, from dormant seeds. The cattails sprouted in the shallows with speed that suggested the vigor of their summer lives. Arrowhead and bulrushes were heading for the surface, and all the plants that would float on the surface had begun their climb there.

Pondweeds and duckweeds and water lilies were impelled upward with an urgency that seemed to reflect the fact that the earliest plants to surface were the ones most likely to survive. Many plants were to live with almost the same intensity and direction as animals and insects. Their resemblance to animals would be best exemplified by some strange plants now appearing as shoots deep underwater, which would spend the summer hunting and eating animals and insects, and other plants.

This was the awakening in the pond, a prodigious expansion of life, diversified and interrelated. In it there was such a common animation of spirit and objective that plants and animals would soon merge insensibly into the secret of the pond and the swarming days that now lay just ahead.

CHAPTER 5

a race in the sun

Soon after one sunrise, the pond was a strange phantasm of itself. The misted ocher sun reached through a sparsity of trees and put its mark on the water with a perfect inverted reflection. As it stretched yellow fingers across the shining flat water, a faint mist rose through it, in places obscuring the rich greens and browns of the pond's banks and giving ephemeral substance to the shafts of the sun. The sun rose higher, the shafts disappeared, and soon the mist had gone and the pond was opening up like a great eye, with the green of the surrounding trees leaping into view as the sun rose beyond them. This was obviously a new season.

The deceptive countenance of the pond concealed a titanic balancing of life. At every level of existence, constant competition for space and a share of sunlight, oxygen, and moisture made haphazard the lives of all creatures. The pond weeds pushed against each other for space in the shallows: the ones that reached the surface early would survive. The symmetrical lily pads stopped the sun from reaching into the water and so discouraged other plants from following them. The duck-weeds were packing their leaves into place throughout the pond, and they prevented most other aquatic plants from getting a share of the surface of the water wherever they grew. Some plants, like cattails, strangled *all* competition. Tree leaves and needles spread in the air to catch as much of the sunlight as possible, making it difficult or impossible for new trees to sprout under them. Despite the pond's show of abundance, its resources were strictly limited. There was plenty of air but not an infinite supply of water in it. There was plenty of water in places, but not an endless supply of warmth and oxygen. There was plenty of sun, but never could it shine on all things equally. All the living things in the pond were racing to get a dominating share of the resources.

There was little outward sign of any struggle. The woods were fragrant from spring flower perfumes. In the swamp, the sun shone on the gold petals of marsh marigolds. Bell-shaped flowers of wild ginger unfolded. The rich odor of flowering arbutus mingled with the various scents of anemone and wild strawberry. The wind stirred foliage softly.

The only noticeable struggle was one of sound. The frogs spread through the pond sent their calls rising and falling in puffs of wind. On northern slopes leopard and pickerel frogs called from among the trees, and tiny tree frogs, perched on branches, sent out slender calls. Frogs bulged their voice sacs and called audibly underwater while clinging to plants. The bass growl of the bullfrog sounded against the piping squeaks of spring peepers. The frog sounds mingled with the uproar from the redwings who had established themselves along the margin of the marsh. These dominant sounds overwhelmed the sweet singing of sparrows and the twittering of swallows lancing through blue air.

One midmorning, a recently arrived loggerhead shrike swiftly impaled a wriggling leopard frog on the thorn of a pondside bush. As the thorn passed through the frog, she screamed once, and instantly all the frogs hushed. For a moment the pond seemed silent, but then the sweet singing of a score of birds became audible.

II

For some creatures, the urge was to get *into* the pond. Water striders, frogs, salamanders, beetles, and others headed for the pond from the forest. During two days, a thousand toads dug themselves out of the ground and leaped through grass and brush to the pond. But for most creatures, it was a race to get *out* of the pond. A big dragonfly nymph shot forward by means of a pulse of water from her abdomen. She began climbing a water plantain stem slowly and clumsily. The light grew steadily brighter as she moved up through the underwater leaves; without pausing she pushed her masked face through the water skin and into the air. Inside her body, a radical transformation was reaching its climax. The nymph's underwater body was, at the point of leaving the water, still functioning, breathing in water and expelling it after extracting the oxygen. But these functions were now being suppressed by the insect's new, adult self, which filled the nymph's tough skin.

She paused in the sun, and the process inside her body quickened. The sun dried the glistening exterior sheathing, and this stiffened. Its strong feet clung to the fiber of the stem while inside it a concealed life strained for release. The tough sheathing split up the back, and a strangely hideous creature emerged, wet and slimy, with primeval wings plastered back against its vulnerable yellowish body. It rose steadily from the slit into the sun, its muscles agonizingly curving its body backwards. The new body dried quickly, the wings wrinkled and stretched, and blood pumped through unseen vessels to flare out the wings widely. The body of the new creature darkened and the network of veins in the wings shimmered. The dragonfly flashed across the pond, leaving the brown hard shell fastened to the water plantain stem as evidence of a transmutation that extended deeply back into earth history.

In one day, two hundred dragonflies left their marine cases: the pond suddenly became an arena of flashing wings. The dragonflies hovered in groups, feeling the power of their new muscles and the scope of their new environment. They were to have a formidable effect on the pond. A tiger swallowtail butterfly, looking enormous in his yellow and black finery, came zigzagging across the pond and in one instant was struck by a dragonfly, who, one day before, had been a submarine creature. Sharp teeth sunk into the butterfly's slender body, and the two creatures fell in a flapping mass to a lily leaf. The butterfly jerked convulsively as part of his abdomen was eaten out. The dragonfly methodically ate the body, then hawked back into the air, leaving the broad wings on the leaf. But as he hovered near the cattails, he was being watched by a creature underwater. A great head shot up, a prehensile tongue flicked out, dabbed the dragonfly, and disappeared underwater. The bullfrog caught three more dragonflies that afternoon, all of them recently emerged from the water and all, perhaps, less cautious because they had recently eaten. He seized a fourth, but the dragonfly wrenched the front of his body free and the bullfrog swallowed only the abdomen. The dragonfly flew across the pond, catching insects as he went, unheeding the imminence of death.

For some creatures, the journey out of the pond was crucial. The May flies—insects of delicate refinement, with wings emerging erectly from their slender, curving bodies—had become especially modified to maintain their numbers around the pond. Countless May fly larvae were ready to leave the water where they had lived for a year or more. Now, each with six legs working, they headed for the surface by climbing up the stalks of lilies and cattails, bulrushes and hornworts, star grasses and smartweeds. They paused in the shade of pickerelweed leaves, in plantain and arrowhead and bur reed and pitcher plants. Their incredible numbers augured a spectacle at the pond. They waited as the sun moved around them, then began shucking off their larval skins. They squeezed into the sunlight, showing the tracery of veins in their wings and long trailing caudal tail plumes. But instead of flying, they waited. Soon they began shedding another skin, this time a fine transparent sheath, and in this last transformation their delicacy was ultimately refined. They rose into the liquid air in the last blush of the setting sun.

The red sun was muted to yellow, shining through their massed wings, and they rose to surround and obscure thousands of specks of red—midges—which were jogging up and down rhythmically while maintaining exact positions over the pond. But whereas the midges were merely disporting themselves, the May flies were flying against time. They were too vulnerable to survive as flying insects, and their emergence, instead of being a triumphant beginning of life, was an anticlimactic end. They would mate and die without ever attempting to eat.

In vivid comparison was the emergence of the young mosquitoes. The mosquito larvae, already hatched from floating egg rafts, had been pushed to maturity by the growing warmth of the pond. They had pupated in the water, changing into tiny creatures that had two breathing tubes piercing their thoraxes instead of their abdomens as in the larvae. They clustered on the surface of small pools and patches of still water surrounding the pond, and they were aware enough to dive from menacing shapes appearing over them. Inside each creature was an adult mosquito. They began leaving the water

one morning: hundreds pushed out of their pupating cases. As mosquitoes, they forced their way into open air and balanced on their floating pupae cases. A hundred of them soon stood in one group, turning in the bright sunlight and stretching their wings to get the most drying from the light breeze. One by one, they rose and floated over the water in search of a first meal. The males sought the juices of plants and leaves; the females would find sustenance in the blood of animals and birds.

The time of emergence was always critical and sometimes disastrous. Hundreds of dragonflies, May flies, and mosquitoes were seized by birds while still purblind from their radical water-to-air transition. Survival during the first moments of life was always haphazard.

During the winter, the egg cases of one of the pond's most peculiar insects, the winged, green, predatory praying mantis, had turned black, horny, and tough. They were fastened to twigs and dead stalks and were almost invisible. Inside each case, narrow air shafts had ventilated the eggs and would soon become escape hatches. The egg cases became soft and spongy in the spring humidity, absorbing instead of throwing off moisture. In one case at pondside the eggs began hatching in the darkness of the case's inner chambers, revealing perfect miniature mantises with large brown eyes and pale yellow bodies. They spilled down the ventilating corridors, attached to each other by long cords of silk. They appeared in the sun and hung, struggling to free themselves from the silk and each other.

By chance, an ant was climbing the stem as the mantises began appearing. Like all ants, he was nearly blind, but he was so close to the case that he could easily see the mantises, and his feelers waved comprehendingly. He ran down the stem while a soft wind and sun dried out the newborn mantises. They climbed over one another, some struggling up the nest case, others dropping to the ground. As fast as the newborn mantises freed themselves, they oozed from the fount. They were bent on emerging as quickly as possible. They could not know that the ant was heading a column of his fellows back through the undergrowth, to the bush, and up the main stem. In a moment the ants were swarming over the mantises, biting into soft flesh, and ripping out chunks of it. Ironically, in only one more day the outer shells of the young mantises would have hardened so that no ant could have eaten them. It was only for this moment that they were so fatally vulnerable to the ants. Many of the attackers dropped to the ground clutching mantises and scarcely paused in their eating when they hit the earth. But the mantises emerged faster than they could be eaten. Many escaped, bitten perhaps, or even partially eaten, but that might not matter for creatures so tough. Finally, the cornucopia was nearly spent. The ants explored the complex structure of the nest, and then most of them descended to earth and disappeared. A few remained on guard at the case, waiting for lately hatched mantises to come down to them.

III

Procreation at the pond was sending a flow of fertility by every conceivable route from male to female. With a thump and a bump, the snowshoe hare—who in his gray and brown coat was now a varied hare—mated with a female, and they ran madly along the edge of

75

the pond. The female raccoons, minks, skunks, and foxes were already pregnant from sexual forays in the snow. The male weasels and muskrats ripped and tore their kind in fights over females. Crayfish fought and wrenched off opponents' claws, but they would grow new limbs. The old muskrat chased a one-year-old rival from the pond and then copulated with a young female. The bumblebees, wasps, and hornets were fertilized from previous seasons. Some ant queens were fertilized from a dozen years before. The hermaphroditic snails sought each other out in the shallows of the pond with rippling waves of muscular contractions that, running along their bodies from back to front, oozed them over the mud. Their beady black eyes at the bases of their sensitive, stalklike tentacles looked for other snails, and their breathing channels pulsed just below the rim of their shells. One snail slid alongside another and inserted his male organ into the female organ of the snail at his side, who in turn inserted a male organ into another snail alongside her—him—and within moments there were several snails copulating passively in a group.

Some animals performed the sex act faster than the eye could follow. Others spent hours, even days, at it. A male pickerel frog mounted his mate among the thick cattails on the north bank of the pond, gripping her with almost convulsive strength, his foreleg fingers thrust into soft flesh behind her arms. He was not then copulating, but his presence sped the development of ripe eggs inside her body. Early the following morning she ejected them in a jellied mass. Immediately the male released his sperm cells over the floating eggs, and the two frogs fell apart from their long embrace.

The physical structure of the frogs made direct impregnation impossible, a disability shared by the salamanders. On a cool night one of them—a male red newt —seized his mate underwater. He held her with his feet and writhed his body violently while occasionally pushing his head against hers. His tail lashed and locked with hers, and the two newts rolled over the mud in clouds of disturbed silt. This was not copulation but courtship. Shortly after dawn the two creatures separated, the male to deposit his sperm cells on submerged leaves, the female to wait for him to finish, and then to suck the cells into her body through her cloaca, the all-purpose vent at the base of her tail. The male cells would wait inside her body till her eggs were ready for fertilization, perhaps immediately, perhaps after some delay.

A pair of spotted turtles moved across the bottom of the pond, clinging together so tightly they looked like one animal. The male had gripped the female's shell with legs and beak and was coasting on her back as she swam, sometimes slowly, sometimes urgently when she ran short of air. He was waiting for the right moment to inject his sperm into her reproductive passages.

A pair of butterflies copulated on a leaf, the male's wings vibrating, the female with her wings spread out, their bodies arched to meet in a strain of fulfillment. His sperm would move to an internal sac and later fertilize eggs moving down the female's oviduct.

The floating mosquito egg rafts were being created continuously. The pond harbored a multitude of mosquitoes who hid in thick bushes, under leaves, and in grass, and who shortly before dusk were sent into the air by some combination of temperature, humidity, and lighting. Collectively, millions of them—all males —rose over pond and marsh. They coalesced into columns that stood out clearly against the moonlit sky, and the sound of their massed wings penetrated dense clumps of leaves and thickets and dusky grasses and the earth itself; and the waiting females rose to meet them. The females disappeared into the multitude and then fell away, gripped by copulating males. They would lay their rafts of eggs in a moment.

The May flies had finished copulating in the meantime. The females dived to the water to lay eggs and the males fell, dying, their wings falling off, their brief lives ended in a few moments of activity over the pond. By midnight, the air was clear of mosquitoes and May flies. The pond looked to another day and to the rise of another horde of insects, now waiting in the greenery or underwater for the supreme moment of their lives.

IV

On an elm branch concealed by new leaves, a pair of yellow warblers prepared to mate. The male advanced along the branch, rising and falling with fluttering wings and puffing out his chest and stretching his body into the shape of a tear drop. Suddenly he flew at her. At first, it seemed their mating would be difficult. He balanced uncertainly on her back. In that position, he seemed unlikely to reach her sex organ, which was underneath and facing away from his own.

As he fluttered for position, the female suddenly twisted her tail to one side. The male thrust his body down and with a tremor completed the act. With perfect accuracy, the tiny tip of his stamen had thrust through her feathers and found a vent smaller than a bird's eye and had pierced it and released a tiny blob of sperm. He fell away and flew into a nearby hickory.

Many of the pond creatures preceded copulation with complex dancing and displays. Salamanders gyrated before their mates under water. The May flies seemed to dance. A leopard frog croaked on a tussock of grass surrounded by a circle of females. A woodcock, a slender-beaked bird with big eyes, flew over the pond in a blur of movement and left a sibilant cry behind him. In the higher air, he would meet other woodcocks and join them in wild flight. One of them would disappear toward the pond to hover like a hummingbird, twisting wildly from side to side before sinking smoothly to a clearing near the water. A female would be watching him somewhere, and he would mate with her. The red-tailed hawk won a mate from the stream of migrant hawks and impregnated her near the great lake.

Distinctive in this polychromatic flow of life were about a dozen ruby-throated hummingbirds who had been arriving at the pond over a span of twenty days. There they came to the peak of their breeding condition. The males flew horizontally sideways, sliding through

the air like leaves running down a fast stream. They
flew vertically, sometimes ascending high over the pond
while beak to beak with a female. They fell to the water
and darted at precise right angles to the shore. As they
demonstrated to the females, their invisible wings
hummed like bees. Their mating was a dainty marvel.

During the coalition of migrants with the pond's
life, the purple martins and swallows elegantly swept
over the water, snapping up insects, dipping their beaks
into the water in mid-flight, and drinking. The martins
congregated to breed, as was their habit, in a hole-filled
bank overlooking the swamp west of the pond. Their
violet heads and backs and their sweeping pointed
wings gave them a sureness and boldness of aspect not
matched by other swallows. The tree swallows, smaller,
green-headed and green-backed, would bank in groups
over the pond—suddenly all metallic green, then, as
they turned, all white. The almost constant twittering
of the swallows communicated their zest. One evening
a pair of them flew quickly over the water, separated
at the swamp, and turned back toward the pond and
met there. They mounted high into the air and touched,
barely visible against the magenta sky. Beak to beak
and copulating, wings fluttering silently, they descended
toward the water. They parted a moment before touching it and darted up over the trees and disappeared
into the glow of the sunken sun.

Every creature was under pressure to mate as early as
possible, and this urge could transcend even the instinct to survive.

A male towhee, black-headed, brown-sided, white-bellied, as big as a robin, pecked at a female, and she flew a short distance. He followed leisurely, and she took sudden flight. Quickly the game turned into a chase. The female sped across the pond, hurtled over the granite rocks, and swooped into the sheltering trees, with the male close behind her. She turned again, and the trees were a wheeling blur as she flew along the southern fringe of the pond. She was becoming excited, and her amatory juices were flowing fast. The male was overtaking her, following her flight exactly, when she suddenly swerved into the forest. She heard the rasping burr of air as his wings swung him behind her. The supremacy of this moment blinded caution. Neither bird saw an intent watcher, a sharp-shinned hawk, a forest hunter of speed and silence who now peered at the ecstatic madness of the towhees with red-eyed concentration. Many times he had eaten when two male birds, fighting for territory or for females, had become heedless of the fact that death was the supreme adversary. The hawk glided behind leaves to a point where he might intercept the towhees. He heard wings thrashing through the trees, and the birds shot past him out of reach. Uneasily, he moved higher into the tree and saw the towhees whirling tightly. The female was making her last turn. She would go to ground in grass on the south bank of the pond. The male towhee could only see her striving wings and her flared tail as she turned. Had his attention been normal, he could have seen the hawk. Then the female was down in the grass and he had seized the nape of her neck. She was struggling, but her tail was angled to one side. He thrust himself down, and then the female, the grass, the trees, and the pond disappeared into a red blur as his last breath croaked through the blood in his throat.

V

In cool water, a leopard frog was suspended for a moment in a seeming void of aquamarine. His yellow-rimmed eyes sought the sparkling sheet of the water skin above him, the filtered sunlight sinking past him, and specks of minute life moving indistinctly before him. An amoeba, almost invisible to him, moved slowly past one eye, its single-celled body pushing out protuberances, or pseudopods, and withdrawing them. Now the amoeba was a circle, now an elongated figure, now briefly multiarmed, or triangular, but always moving.

The frog ignored it and kicked languidly to the surface and to the murmuring, scented upper world. The swirl of his passage sent the amoeba and a thousand of his kind spinning through the water. These single-celled organisms now numbered billions in the pond, and they were infinitely more varied than the visible creatures. To many other pond animals, the amoeba was a giant, as big to them as the frog was to it. The sun sparkled through a mass of fine granules, or droplets, in its body, each refracting the light, each moving sinuously as the jellied creature lunged through the water in a ceaseless hunt for food.

On one part of its asymmetrical form was a large orifice, or vacuole, which contracted and expanded like a mouth. The green sun shone uncannily through the pulsing protoplasm around it and revealed the bodies of many small creatures, mostly flagellates, being digested in this and other, smaller vacuoles nearby. The flagellates were both animal- and plantlike creatures not dissimilar to the amoeba, but all were much smaller and had trailing, lashing tails, or flagella, which spun them through the water. One of them, a euglena, about one tenth the size of the amoeba, approached on a course of near collision. As the euglena was about to pass, the amoeba reached out a pseudopod that quickly nipped around the creature and maneuvered it into a food vacuole. Digestion began immediately. Before the amoeba had swum the width of a bullfrog's eye, enzymes were flooding into the food vacuole, killing and digesting the euglena. By dusk, its body had disappeared.

As the water reddened in the dawn of the next day, the amoeba was swelling, having ingested many more animal and plant victims. When the vacuoles had digested the last of them, the amoeba grew slightly larger and suddenly changed strangely. It rolled into a tight ball, its arms stilled, and its dark nucleus—a blob of mysterious animation inside it—stretched and distorted. Then it ruptured into two separate nuclei. Once the internal division was complete, the amoeba's body divided, and the two equal droplets of protoplasm, each containing half the ruptured nucleus, drew apart. The division was unseen. None of this protozoan life had vision, though some of it had a little sensitivity to light, which could send the creatures rising and falling in the water as the sun appeared and disappeared during days patched with clouds.

The divisions of the amoebae occurred an immeasurable number of times a day. Other protozoans, like the paramecium, a slender, rapid-swimming creature whose body was fringed with waving hairs, were as fecund but seemed a step ahead in evolution. Two came together and exchanged fertilizing material. They took one eighth of a day to divide into four new animals, which would themselves be dividing within a day or so. Unchecked, the paramecia would soon fill the pond.

The paramecia and amoebae and other divisible animals had evanescent life spans. The creation and destruction of this microlife was prodigal: animals and plants appeared and disappeared in billions. The intricately shelled diatoms spread up bur reed and water plantain stems, and teemed on the surface, on rocks and on larger creatures. A dragonfly nymph crawled up a stalk into the air, carrying with him thousands of diatoms stuck to his body. They died in the sun as the flying creature left his marine body.

The desmids proliferated equally rapidly in starred, rectangular, half-mooned, and square shapes, and also spread over rocks as green slime and collected on the backs of swimming nymphs. Their numerical explosion was so gigantic that the pond turned green, yet no individual desmid was visible anywhere.

In this great upsurge of life, the hard-shelled diatoms had developed one ingenious method of surviving in spite of a seemingly insuperable difficulty of their lives. Each diatom consisted of two separate sections of shell, one of which fitted into the other. When a diatom was ready to divide itself, the protoplasm inside the creature shoved the two shells apart, and each division took one shell with it. Both creatures then grew a new shell to compensate for the one lost. But the diatom that had inherited the smaller shell would, when it divided again, produce an even smaller shell, and this process of reduction went on till further division was not feasible. The diatom protoplasm would then ooze out of the shells, seek a full-sized diatom and merge with its protoplasmic contents.

The differences between plant and animal among this microlife were incomprehensible and contradictory and seemed to indicate only one fact: the origins of this life went back to a creature that was neither plant nor animal.

The plantlike euglenas, lashing their flagella with animal-like purpose, were reproducing in vast numbers in the pond's southern shallows. They contracted and expanded like amoebae, but with more precision and energy. When swimming, they rolled into spheres and then expanded suddenly into long slender shapes, which went jerking through the water. They were insensitive to the passage of diving beetles and salamanders which swept through their colonies, and they zigzagged about in the sun, creating their food through photosynthesis as perfect and self-contained microcosms.

The growth of this invisible life in the pond had only begun and was increasing its speed. As the green algae spread over rock and plant stem, they attracted hundreds of snails, who grazed on their masses and in turn were preyed on by increasing numbers of birds and animals. The population of the pond grew vastly as the sun poured down and generated more infinitesimal life, which carried on the infinite process of sustaining other, more complex, lives.

CHAPTER 6

the hunt

The pond swarmed with creatures whose lives depended on guile, dissimulation, stealth, on strength and speed. The hunters ranged the pond from the high upper air, where the red-tailed hawk hovered, to the mud bottom, where dragonfly larvae lurked in endless ambush.

The death of the towhees in the southern forest was an insignificant incident in one hunting day. Before dusk the sharp-shinned hawk had killed again several times. But the towhees' death exemplified the perfection attained by some hunters in securing their prey.

Among these creatures were hawks, owls, wasps, robber flies, minks, weasels, mantises, and crows. Perhaps the most imposing was the red-tailed hawk. His scream in the hot sky gripped the senses, and the vertical fall of his body terrified the forest. Another was the great horned owl whose booming night call jerked sleeping crows out of their sleep. If crows dreamed, they would have visions of great horned owls.

But size was no criterion of the hunter's skill and power. Few creatures better personified the hunt than the robber fly. One had appeared now, and his arrival had been so fleet that he was still unseen at the pond. He was black, slender-bodied, hunch-shouldered, with enormous multifaceted eyes and long hooked legs. He stayed unmoving on a cattail stem, his eyes glistening and watching. The pond sparkled. The trees murmured in the wind of a new season. The frogs were silent at midday.

The fly was motionless one moment and gone the next. Only the acutest eye could follow him low across the water on a course of interception that ended at a droning bumblebee. The impact flung the bee into the grass. He squirmed to sting the fly, but the powerful grip of hooked legs held him tightly till a sting jabbed into his back. He was still struggling feebly when the fly inserted a sucking beak and began withdrawing his body fluids.

Meanwhile, another robber fly arrived, flashed over the feeding fly at pondside, and disappeared toward the northern trees. She saw a large beetle scuttling over some bare gound and dropped and straddled him before he even saw her. The beetle ran while the fly pushed her sting at his heavy armor, seeking to find the chink that made him fatally vulnerable. She found it, and as the beetle slowed his scuttling run she unsheathed a cutter from her mouth and sawed a hole in his back. Through this she inserted her beak and sucked. The beetle quivered and died.

It was a time of hunters lurking behind the peaceful façade of the pond. A yellowthroat jerked out his song in a thicket. A volley of crow calls rose from the marsh. A big green darner dragonfly hovered over the feeding robber fly for a moment, then flicked over some blossoming chokeberries to the pond. He lowered a scoop

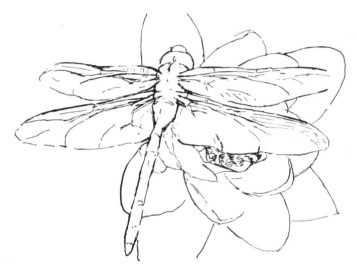

formed by his hair-fringed front legs and drove through a swarm of gnats flying close to the surface of the water. A dozen of them were imprisoned in the scoop, and while still flying, the dragonfly chewed them dry and dropped their bodies in the water.

As long as a dragonfly patrolled the pond, no insect was safe. Some of the hundreds of dragonflies there were as big as small birds, and all had double wings outstretched flatly from their bodies. All had great compound eyes that were almost omnipercipient, so widely and minutely did they see. The big green darner saw a wolf spider running through grass on the pond bank. At the same time and in the opposite direction, he was watching a monarch butterfly resting on a twig. The dragonfly's head was flexibly joined to his body so he could drop his head and look underneath and behind, and fly, as a result, with exceptional precision. He might be halfway across the pond and flick himself one hundred wingspans into the path of a butterfly. He would be eating before landing with his struggling victim.

The dragonflies were catholic hunters and cannibals. The green darner paused over a leaf, then struck at a smaller, bright red dragonfly nearby. But his prey eluded him. He flew backwards, then whisked off sharply and hit a low-flying wasp. The two creatures tumbled into a mass of aquatic plants, the stingless dragonfly trying to kill the wasp before she could drive in her sting. The insects disappeared into the greenery, and the bright sun glittered on the tranquil pond.

The wasps greatly outnumbered the dragonflies at the pond and were their antithesis. They were members of a great family of earth insects who had evolved lives of refinement and ingenuity, in contrast to the dragonflies' simple attributes of speed, power, and rapacity, which were virtually unchanged since primeval times. Some wasps, not seen at the pond, used small pieces of wood as tools to dig their burrows. Others, at the pond, were knowing enough to wrestle armor-sheathed victims till their bodies bent and a small crack appeared. The wasps would then sting through it. All the wasps were shelter builders, and they were both solitary and communal.

The silent struggle between wasp and dragonfly continued in the greenery; nearby, six different wasps lingered on a cluster of lily pads. All had come to drink. A solitary wasp, black and skinny, his body longer than his wings and his head joined to his body by a short, exceptionally thin neck, rested on one leaf. He looked light and fragile, and a puff of wind moved him fractionally. Near him was a cicada killer three or four times his size, with massive abdomen, thorax, head, eyes, and wings. As a hunter of creatures equaling or exceeding her size and weight, she needed vigorous strength. Her shiny black body was half-circled by yellow bands, and she seemed remote, in the evolutionary process, from the fragile solitary wasp.

Elsewhere, communal wasps were working. The sun flickered over the iridescent coat colors of the purple mud daubers, making them blue one instant, then green or purple. Potter wasps—shiny black and yellow, delicate and industrious—gathered clay to make small spherical nests. Paper wasps, with yellow stripes over green abdomens and red spots decorating their sides, were masticating wood fiber to make a nest. The puzzle of the wasps was why some collected into communities and fed their youngsters on nectar while others were solitary and fed their larvae on red meat.

The robber fly was still feeding on the beetle when a

hunting wasp hummed through the whispering green of oak and hickory. She had seen the shiny brown face mask of a large caterpillar bobbing up and down as he chewed on a leaf, his body invisible against it. The wasp alighted on the leaf, and the caterpillar tautened into a ball and dropped. Down he fell, glancing from leaf to leaf and then hitting a branch and ricochetting into the apparent safety of thick grass.

The wasp sank slowly after him. She found his rolled-up body wedged between two grass stems, and with deliberation, she stung him twice. The stings might kill him or they might merely paralyze; it did not matter. His meat would still be edible when the wasp's egg hatched inside his body later. She gripped him now and pulled him free from the grass. In a moment she was in the air with him, headed for her subterranean burrow and the laying of one egg.

A yellow-banded wasp rose slowly and awkwardly from the pond greenery. She had been severely bitten and partially blinded by the dragonfly, but she had killed him at last, driving her sting through a chink in the cluster of his six legs. The poison paralyzed the dragonfly's simple organs and stopped his long tubular heart. The wasp resumed her interrupted journey.

II

Many moments of pure terror occurred in the hunt: a violent flurry of forest leaves, a sudden impact high in the trees, a long, squalling cry of anguish, savage fights underground and underwater.

A squirrel, walleyed in terror, raced through the undergrowth towards a tree, with a weasel close behind her. An owl dug into a patch of leaves in search of mice, and they squealed when they found their exit tunnels collapsed. The squirrel might escape, and the mice might die; both were trivial events. Every day brought new variations in the hunt.

A gray fox's narrow, knowing face looked over the pond one morning and sent a squirrel leaping into a tree, from where he saw the fox climbing after him. The squirrel chattered in anger and anxiety and jumped higher up his now dubious stronghold, but the fox came on inexorably, choosing his footholds carefully, digging sharp claws into bark to hoist himself up the branchless parts of the trunk. He was using an innate and seldom-practiced faculty to climb, but he was hungry. Soon both creatures were high over the pond: the squirrel could see the gleaming marsh. He ran to the end of a long branch, and the fox cut off his escape by mounting the branch at its base.

The squirrel could not jump to any nearby branch. The ground lay without obstruction far below. The fox edged along the branch, the near-hypnotic fixity of his eyes inviting the squirrel to run back past him. But the squirrel met this unexpected crisis with a rarely used capability; he launched himself into mid-air, so high he looked like a falling bird, his body flat, his four legs spread out as widely as possible, his fluffy tail bushed to catch the uprushing air. He thumped soggily into grass and mud and bounced away into the forest.

One morning, a white-bellied, fish-hunting hawk—an osprey—arrived at the pond from the marsh and saw a dace. He fell under rigid, upthrust wings, his breast arched and his talons thrust ahead of him. For the fish, the arrival of the bird came in a bubbling explosion of river water and a violent blow of claws striking. The osprey shrugged to the surface and lifted out of the pond in a powerful sweep and swirl of mottled pinions.

The squirrel watched the osprey fly to the marsh and heard the dace flapping, but now it was his time to fall prey to the hunters. The red-tailed hawk had been watching him. He had dropped quietly down into the forest with his mate, and the two birds converged on the squirrel's tree to block his escape and confuse him. With the female hawk beating through the leaves after him, the squirrel raced into the topmost branches and launched himself again into space. But the male red-tailed hawk took him easily, and the scream of terror, mixed perhaps with surprise, was cut off in a clench of claws.

The winged hunters, though formidable, were not invincible. Many ground creatures, like minks, weasels, and otters, would fight to repulse attacks from the sky and sometimes even would attack the flying hunters. A hawk eating a duck in the marsh was seized in the neck by a weasel, and the two animals started into the air before sharp teeth severed the hawk's central nerve system and he plummeted into the water. The weasel swam from his body to begin eating the duck.

The hunters could easily misjudge the dynamic instinct to live of their defenseless prey. The varied hare crouched into grass when a Cooper's hawk arrived at the pond one morning. The hare looked at the hawk's orange-striped chest, his lethally shaped body, saw the sheen of sun on his blue back, saw his bright red eyes. The hare did not separate this hawk in his mind from others he had seen hunting at the pond, but he may have been dimly aware that this one was particularly dangerous. He had once seen this hawk and a mallard smash together over the pond in a shower of feathers. The birds, locked together, had fallen to the water, and the terrified duck had managed to flip beneath the surface, carrying the hawk with him. For long moments, the surprised hawk contested the submarine struggle in an environment that was totally repugnant to him. Finally, he had burst, choking, to the surface and with a clash of wet wings rose from the pond. Despite this setback, he had waited in a nearby tree for the mallard to reappear, not knowing that the mortally injured bird had drowned in a constriction of aquatic plants.

On another occasion, when concealed in the grass, the hare had gone rigid as the same hawk walked past. The bird moved to some thick grass on a low ridge and peered through its stems. Then he backed off and with a powerful thrust of blunt wings shot out of sight over the grasses. The hare had not seen the hawk when he had been flying toward the pond and had glimpsed a grazing cottontail rabbit through the trees. The hawk had stolen up on the animal, flying between thickets and bushes. The hare only saw the hawk's final stalk on foot and his sudden flight of attack.

The eyes of the two animals now met in the sun. The hawk swiveled, and as the hare shot away he got one last glimpse of the looming hunter and knew he was lost. At the start of his third giant leap, he felt fire burning into his rib cage. But the hawk had misjudged the hare's speed. Instead of gripping his head and ending the hunt with a clutch into the brain, he found himself holding loose fur and flesh and riding the leaping animal in surprise and fury. The situation was beyond his, and probably any other hawk's, comprehension; in six leaps, the hunt was over. The hawk was knocked sprawling as the hare ran under a fallen branch, and the animals parted, the hawk to preen his disarranged feathers in a tree, the hare to disappear, panting, into the forest, and lick his bleeding wounds and endure sickening pain seeping into his body. His predicament, of uncertain outcome, was typical of the hunted who by temperament were especially fitted to silently bear their injuries in thickets and burrows while the hunters went on to attempt new kills.

But at rare times, the hunter might have to bear severe injury himself, and his mien then would show the great gulf that separated such creatures from their prey. At dusk, the hare sniffed a sour smell of corruption above him and crouched. A great horned owl had arrived silently and unseen, and the odor he brought with him came mostly from the small head of a mink that was fastened to his back. Days before, the mink had attacked him at the marsh while he was pulling at the remains of a freshly killed heron. The mink had bitten him several times on the back before locking her teeth in a deep grip that the owl could not shake free. During the fight, he had turned his face to his attacker and gouged out one of her eyes, but he could not break the grip. Finally, he had killed her and nipped her spinal column through, but he still could not tear the head loose. He soon came to ignore the head, which in rotting would drop off.

The corruption troubled him not at all. His feathers were so often spattered by last explosions of life in his victims that he reeked permanently of death, a smell mixed with the odors of musk from muskrat, weasel and mink, and, overpoweringly, the stench of skunk, the flesh of which was one of his favorite foods. He flew over the pond with the mink head bobbing on his back. The hare licked his throbbing wounds and trembled.

III

The hunt led inflexibly to death. The real hunters knew no fear. One evening a young raccoon saw, for the first time, a praying mantis on a leaf in some alders. He was fascinated by the insect's strange stance, the four supporting legs holding his wing-sheathed green body parallel to the leaf, his forebody angling up sharply, almost vertically and his neck jutting forward. His red eyes, with their wide scope of sight, saw ahead, up, behind, and down. Underneath his rearing head sprouted

97

his spined front legs, jointed thrice, held in relaxed readiness for a lightning movement that could capture any flying insect at the pond. He watched the raccoon intently. The raccoon reached forward to touch the mantis, but the insect jerked menacingly, his spined legs flexed for attack and his antennae waving furiously. The raccoon was impressed but still inquisitive. He reached out again, and the mantis struck his paw. The two creatures faced each other in the dusk, the animal still doubtful but curious, the mantis a cold and relentless resister of all his intentions. Then the raccoon turned and ran into the bushes, and the mantis swiveled his head to keep the animal in sight.

The mantis was the perfect hunter, seeking his prey day and night, capable of enduring long fasts but also of eating heavily when prey was available. By the following morning, he had flown to the pond's south bank and was merged into a background of cattails. On the land side of the cattails, the shoreline was spattered with flowers and through these floated a ruby-throated hummingbird. Her wings were invisible against the early morning sun, and she moved from flower to flower without seeing the mantis. His head was moving slightly, keeping her in view, and his body was rigid. The hummingbird hovered over a bee balm flower, then moved toward a spatterdock, but never reached it. The mantis' powerful legs arced outwards, and the inside spines crushed the tiny bird into a formless ball of feathers and flesh. One of her wings fluttered uselessly. Her pretty colors flashed at the peak of the upthrust position of the mantis. His red eyes disappeared into her feathers as his mouth sought her living flesh; her tiny dark eyes blinked in terror. A water thrush teetered on a nearby log, and a bobolink called distantly. A cluster of green feathers floated slowly among flowering marsh marigolds.

The mantis was one of fifty or so who lived at the pond, and he, or any of his kind, might never catch a hummingbird again. But they ate anything they could catch, except ants. They were almost two insects in one. The lower part of their bodies fastened by four legs to a leaf, twig, or stem, leaving the upper part of the insect free to deal with the struggling victim. The mantises were armor-sheathed in chitin, the horny exterior covering peculiar to insects, and were almost impervious to stings.

But no pond creature was safe in the hunt. Chitin had one fatal flaw. It could not expand with the growing bodies it contained. All the chitinous creatures at the pond had to shuck off their old skins as they grew and emerge in new soft ones which hardened into sheathing. After the mantis had finished eating the hummingbird, dropping feathers, bones, and legs into the grass, his growth quickened toward a skin change.

During this season, he had caught and eaten more than a score of yellow jacket wasps who hunted among flowers along the pond's south bank and had a nest nearby in an old mouse burrow. Many of the surviving yellow jackets had seen the kills, and they were aware of the mantis and his hunting ground. They saw him frequently during their hunting.

In the deep shade of the southern undergrowth one afternoon, the mantis gripped a plant stem firmly in readiness for his skin change. His green back split up the center, and a new naked creature began emerging from the skin. It was identical to the old except that it was yellowish and soft. In that brief moment of emergence, the mantis had been seen. A yellow jacket wasp had watched him fly into the shade and followed him. He watched the mantis free himself from the old chitin. The wasp seemed to know he was helpless. Once the mantis was clear of the old skin, the wasp flew up behind him, settled on his new wings, and stung him twice. The mantis fell, and by late afternoon he was being eaten by a dozen yellow jackets.

When hunter met hunter, the tenacity and toughness of these creatures might be revealed. One night a small screech owl came across the pond. As he glided to a landing, he saw a rigid red-eyed mantis watching him from a twig. The owl changed course slightly and seized the mantis, but his grip was faulty. With a sudden twist, the mantis freed himself—and also wrenched off his own head. The owl dissolved into the gloom, and the mantis flew toward the pond and landed clumsily on an arrowhead leaf. He remained there through the night, his headless body moving spasmodically as though reaching out for a passing, and nonexistent, insect. He maintained his characteristic hunting pose till the middle of the following day, when he fell to earth. But he was still alive. He died while being eaten by ants and beetles and other scavengers at the pond.

IV

The success of much of the hunt depended on the separate personalities of the hunted. Some birds were unwary, always flying tardily from danger, or sleeping in exposed places, and these were killed first. One morning two crows arrived at the pond. They were raising four youngsters, and the demands of the young birds had driven them to extend their hunting territory. Their arrival sent a pair of robins and a pair of jays hastening into the treetops after them, screaming wildly and flying at their heads. The crows moved steadily through the trees, condescending to notice their attackers' presence by crouching at a suddenly bold attack. The leaf-shrouded robins' nest escaped their attention, but the blue jays' nest, now almost overflowing with young birds, was easily visible in a high tree fork.

The male crow descended deliberately and with a single swift jab impaled one youngster. The young birds burst from the nest with their screaming parents striking the crow's head. The female crow dropped under the nest and grabbed a fleeing nestling in mid-flight, decapitated him, swallowed the head whole, then turned sharply and grabbed another youngster floundering in a clump of leaves. The male crow had gone, and the female followed with her beak full of bloody harvest.

The crows, in their anxiety to be off to their youngsters, had overlooked two blue jay chickens. One lay dead and headless in the forest. The other had left the nest as soon as the attack began and, not being well fledged, had dropped rapidly through the leaves, smashing through a low bush at pondside and hitting the water. He remained for an instant with outspread wings, his gulping beak underwater. Even his parents had missed his escape while trying to harry the crows. But there was nothing they could do for him now. He flapped across the water, terrified and confused, moving away from the shore. As the crows left the nest, his efforts were weakening, and he drowned by a lily leaf. His wings spread out and his blue feathers, now darkened by the water, were almost invisible against the dark bottom of the pond.

With the crows gone, the shouting cries of the parent birds died away. The robins returned to feeding their youngsters, and the jays remained in the tree containing their ruined nest. The male bird called harshly; the female flew off and returned with some grubs and worms. Without showing indecision, she ate the food and cleaned her beak on the blood-spattered branch before her. The jays left the pond later that day and were not seen there again.

V

All pond creatures had particular enemies who perpetually haunted their lives. If there were great horned owls in the forest, the skunks lived only from moment to moment, never knowing when one of the big birds would coast in silently and unseen from above. Even the marsh mink were wary of these owls. One mink, seeing a huge owl gliding at her kittens at dusk, leaped high in the air in a desperate effort to stop him. She disconcerted him enough to parry his attack.

Even the fox had an enemy, who was, however, much less dangerous than he was annoying. A great gyrfalcon, who had visited the pond in the late winter, successfully stole food from several fox kills. On one occasion, the big agile bird had exasperated the fox by darting repeatedly at him, always just avoiding his snapping attacks. Eventually, lured too far from the carcass of a half-eaten rabbit, the fox had seen the gyrfalcon rise overhead just out of reach, seize the rabbit, and disappear with it toward the marsh.

Many pond creatures had not one enemy but several. The dog muskrat stood erect on his piled-up mound of leaves and reeds and watched for hawks by day and for owls by night. Above everything, he dreaded the arrival of a lithe, fleet animal ten times his size who could easily swim him down. An otter had visited the pond in the winter and with his long teeth had quickly enlarged the hole in the ice that the muskrat used to leave and enter the pond. Under the ice, the otter had swum to the entrance of the muskrat's underwater shelter and had tentatively inserted his sleek blunt head. Then, rising to the ice, he had exhaled a bubble of air, which he had pushed along under the ice with his snout, perhaps trying to purify it. Then he had sucked the air back into his lungs and disappeared into the gloom.

The otter was the muskrat's worst single enemy; the mink were multiple enemies, slipping into the pond in twos or threes in search of him and his family. On several occasions, the muskrat had fled to a deep-dug hole in a pond bank. One mink dared to enter the Stygian tunnel, and in clouds of mud the muskrat had fought for his life. In that restricted space, his long front incisors were redoubtable, and the bleeding mink retreated.

The cottontail rabbits, who had occupied several deep burrows on the pond's north bank, had enemies all round them, but they were most fearful of the weasels who, silently as snow, entered their burrows at night. Some of the rabbits were lucky enough to have escape tunnels, dug by the long-departed excavators of the burrows, and they fled through these the instant they had a whiff of weasel scent. But in the darkness of their terror, they sometimes blundered into each other or into blind tunnels or froze rigidly while the scent grew stronger and the patter of their own hearts sounded in their ears.

Terror was fundamental to the pond. It heightened the acuity of herons in the marsh at night so that they were warned to take wing before the unseen, silent,

stalking mink reached them. It gave ducks momentarily supreme agility to avoid diving hawks and sent the varied hare off in a blur of speed from some unrevealed night hunter. The hunt went on perpetually, and its victims waited, eyes wide and nostrils flaring, for its arrival.

VI

The pond concealed hunters who could not pursue but who waited in the feeding territory of their victims and killed them while they ate. Brilliant moccasin flowers burst into bloom in the woods surrounding the pond, and chokeberries blossomed thickly around the granite rocks. Painted trilliums and bellworts put color into the shores of the pond. Wild spikenards dangled flowers in zephyrs. An oriole sang, and a butterfly jerked down to a flower. Barely had she settled when her wings splayed grotesquely. The flower swayed, but the butterfly seemed held fast. Her wings flipped and the bees droned around her and later the wings fell from the flower. There was no sign of the butterfly inside the blossom, but the black eyes of an ambush bug glinted in a body of heavy spined back armor. He retreated farther into the flower, moving his massive claws into position for the next attack. Nearby, a bumblebee buzzed frantically as another ambush bug grabbed him. This time the hunter's resources were fully tested. The bee, though ten times as big as the bug, could not break from the claws. His best chance was to pull the bug out of the flower. But the bug's powerful back legs were dug deeply into the petal tissue. He maneuvered the frantic bee into position and stung him between the eyes.

The trees were filled with spring canker moth caterpillars, who were marked with vitreous yellow and black stripes. They moved from leaf to leaf, lowered themselves on silk threads, and hung in mid-air. They constantly ate leaves. Eating *them* were the stink bugs, who were flat, strong-legged, and thickly sheathed in chitin shaped in flat geometrical planes. One was moving back and forth across a leaf, reaching for an unheeding caterpillar feeding on a nearby leaf. For some reason, the bug seemed disinclined to fly, perhaps knowing that landing on the leaf might frighten the caterpillar into falling to earth. Instead, from under the head the bug unfolded a jointed member, which snapped into an extended position in a series of precise motions. He jabbed the caterpillar with this, lifted him up easily, then settled back to eat. When the caterpillar was deflated, his body juices transferred to the hunter, the bug dropped him, folded the hunting apparatus back under his head, and flew higher into the leaves.

The ant lions lived on the peninsula of sand and rock that jutted into the pond. They lurked in patches of sand and depended for most of their sustenance on colonies of red ants who also lived on the peninsula. The ant lions were bristle-bodied, with great barbed pincers and big bulging thoraxes which, when filled with food, could sustain the creatures for days. The lions reached the sand as eggs dropped by double-winged flying creatures, smaller than dragonflies but similar in appearance. When they hatched into ant lions, they dug into the sand, each forming a concave hollow sloping down to their buried, waiting pincers. The sides of the pits were graded to crumble when an ant had only half his feet on the slope. He might turn back, but the ant lion would dig frenziedly, to send the sand, and the ant, cascading to the bottom of the pit. The pincers would close and the ant's body juices would quickly drain into the ant lion. This cycle of hunting was repeated hundreds of times during the summer, with one variation.

A wasp, crippled in a skirmish with a robber fly, fell on

the island and during her stumbling passage over the sand tumbled into an ant-lion pit. The pincers snapped shut instantly, but though the wasp was caught she squirmed into position and stung the ant lion in the face. The wasp dragged herself from the pit but died nearby. The red ants found the ant lion's body the next morning, and they dug it out of the sand. Under the full hot sun of the season the pit dweller was eaten by his prospective victims.

VII

The hunt concentrated in the pond, where billions of creatures preyed on one another. The supreme hunters were visible, like the olivaceous great diving beetle, slipping rapidly through the water in his hunt for tadpoles, snails, large insects, and sometimes, other great diving beetles. He would eat anything he could catch but moved cautiously when he saw an even bigger beetle, the water scavenger, gliding ahead of him. Both beetles resembled marine animals, with their sleek shapes and hair-fringed legs that kicked them through the water. But they were dwarfed and intimidated by a third type of submarine hunter who looked like a land beetle accidentally living underwater. The giant water bug had powerful grasping front legs and tapered, tight-folded wings. He frequently caught and ate young fish.

But none of the water bugs or water beetles were wholly aquatic, like fish. They used the water to hunt and breed in but revealed their terrestrial origins by unfolding wings and flying and by enfolding air under their wings at the surface and breathing it underwater. Some beetles even sucked air out of plant stems.

The water beetles, like those on land, were especially aggressive and successful. The old muskrat, poking in the mud for shellfish and snails, was momentarily distracted by a great diving beetle soaring close overhead.

The beetle curved down toward a large, frantically fleeing tadpole. But in mid-flight, the tadpole was pulled to the bottom by a nonswimming, slim, six-legged diving beetle larva. In another moment, the beetle himself swept up and also grabbed the tadpole and in the rush of his attack dragged both tadpole and larva across the mud.

109

The muskrat remained motionless, bright eyes watching the fight as beetle and larva pulled against each other, but he saw only part of the struggle. He could not see the larva using his mandibles to inject a strong digestive juice into the tadpole, which began breaking down the body tissues into a thick liquid. Unexpectedly, the beetle broke off the struggle and disappeared. Stopping the flow of juice, the larva used his pharynx to suck the predigested mixture into his stomach. The tadpole was soon a skinned husk. The muskrat moved at last, rooted out a mussel and headed for the surface, where he would drop the shellfish in the sun. When it died and its shell gaped open, the muskrat would eat it.

Hunting senses were as acute in the pond as out of it. The water carried sounds and smells. After a large leopard frog tadpole escaped from a dragonfly larva's attack, he swam sluggishly while his clear body fluids seeped from holes made by the larva's mandibles. He hung helplessly in the still water while a horde of tiny predators converged on him. They were planarian worms, and they surrounded the tadpole. One large planarian, a quarter of his size, embraced him and pressed both body extremities hard against him. A tiny pharynx emerged from his mouth in the middle of his underbody and plunged into the tadpole. Other planarians were also fastening themselves. A score of pharynxes sucked the tadpole's body juices into the attackers' stomachs. Both the feeders and the stricken animal sank. When the planarians scattered, the tadpole had disappeared and only a tiny scuff mark in the mud remained to show that the incident had happened.

The tadpole's death occurred near a hunting plant, the bladderwort, many of which grew along the southern shallows of the pond. Their foliage spread gracefully but did not reach above the surface, and their bright green and finely split leaves were fitted with many hair-lined traps that could be pushed inward but would hold tightly against any effort to pull out from them.

The sun streamed down through the water, and the carnivorous bladderworts fed. Tiny crustaceans, worms, and small insect larvae swam into the traps in response to some indescribable attraction, and an aquatic beetle, sliding through the foliage of one plant, passed a thousand creatures being digested behind the fine hairs of the traps. In an hour, one bladderwort caught five hundred thousand creatures. Two hundred bladderworts grew along the south bank, and they ate steadily all day.

Throughout all pond life there was death, and it came in many unexpected ways. The dampness and decay of the swamp near the pond was fostered by the shielding cedars, tamaracks, and spruces, and this environment harbored the hunting pitcher plants, which sent up slender stems and large purplish flowers from broad spreading leaves. The leaves were rounded, hollow, and curved to catch rain water in a pool at their bases.

Each leaf was thickly covered with hairs, which pointed toward the pool and encouraged the downward course of insects seeking to reach the water. A caterpillar who

111

had dropped on a silk rope from a tree moved down one of the leaves now, undulating easily over the tiny hairs. He paused at the water to drink, then turned to move back, but now the hairs were against him and he could not move from the water. Eventually, exhausted by pushing and probing at the hairs, he fell into the water and drowned. His body would decompose there and in time would be digested into the plant.

A small fly glistened in the sun as he flew over some marsh marigolds in the swamp. He saw some widespread hairy leaves that were tipped with numerous minute drops of nectar. He flew down past a tall slender stem bearing a bunch of white flowers and landed on one of the leaves. Instantly, his feet were caught in the clinging nectar, and as he struggled to escape, the hairs around him slowly bent over and trapped him more securely. During that afternoon, the leaf produced a digestive juice that slowly flowed over the now dead fly and the body began to dissolve into the fibers of the carnivorous sundew plant.

The hunters, whether plants or animals, all seemed developed to do the extraordinary, to reach a little further, a little faster. A stricken worm sank slowly through the foliage of a pond bladderwort. It was nearly dead, and its body was encircled by clusters of fine rods. It settled on the muddy bottom and died. Another worm fled through the bladderworts pierced by a rod that had been driven clear through its body. A colony of hydras had fastened their bodies to the underside of the taut water skin and looked like branched plants at first, but the branches were actually tentacles and the hydras, about two hundred of them, were animals and hunters.

The warmth of the sun had attracted great numbers of water creatures to the surface. As water fleas, worms, and crustaceans swam past the hydras, many of them disappeared in a blur, taken by tentacles that moved too swiftly to be seen. The disappearing victims might be glimpsed as they were thrust into the mouths of the hydras near the stemlike bases of their bodies.

The hydras struck at any creature they could grip and frequently seized animals difficult to subdue. Hunter and hunted would sway like plants in a wind. The hydra would then drive in fine, rodlike stings, concealed in lumps at the end of its tentacles, and would grip the victim till it was conquered. There seemed no limit to the appetites of the hydras. Some, during one afternoon, ate thirty or forty creatures amounting to several times their bulk.

The hydras' hunting was disrupted as the sun was edging behind the western willows. Below them, a sunfish darted at aquatic insects among the bladderworts. He was watched from above by a kingfisher who had taken sporadic flights across the pond during the afternoon. The bird dropped into the water suddenly and seized the sunfish, and in the sparkling detonation of water the hydras were scattered, lost their holds on the water skin, and drifted down to the bottom, tentacles writhing.

The greatest hunt was invisible. The billions of protoplasmic rhizopods, with their jelly bodies and blind questing arms, were the hunters of diatoms, desmids, rotifera, and other simple animals and plants. At this level of life, hunting was instinctive, a lunging response to vibrations in the water, perhaps a reaction to smells, to shifting shadows. All these hunters were blind, though some of them had light-sensitive cells. A rotifer came angling down through the green-tinged water, its head cilia a blur of action. A worm came rushing after it, vastly bigger than the rotifer but only barely visible to a nearby pickerel frog. The worm gulped down the rotifer, and the frog swallowed the worm. Later, the kingfisher killed the frog, and the hunt passed endlessly from creature to creature.

A vampyrella, an orange-colored amoebalike creature, drifted through the water and settled on an almost invisible thread of plant tissue. It thrust its body against the thread and pierced it; the plant cell structure changed and suddenly collapsed. The vampyrella sucked out the contents, an action it repeated a dozen times that afternoon.

Another hunter was the actinophrys sol, which had a foamy body from which sprouted numerous hairlike rays, giving it the appearance of a sun shedding light. It moved steadily forward with no visible means of propulsion. When other small creatures collided with the rays, the creatures appeared suddenly paralyzed. One remained motionless on the tip of a ray while a small piece of protoplasm flowed outward along the ray from the body of the actinophrys sol. It enveloped

114

the victim, then guided it down the ray into the jellied body of the hunter.

Though much of this hunting seemed accidentally successful, the invisible organisms actually searched for their prey with much the same purpose as did creatures scores of thousands of times bigger. The tiny paramecia, now teeming in the pond, were blindly hunted by one of the protozoans, a didinium. One of these animals darted forward to butt its long proboscis against the towering bulk of a floating frog, turned back, and nearly rammed an amoeba. Then it collided with a paramecium, and its proboscis instantly punctured the creature. The hunter's body flared open wide and the paramecium disappeared into the didinium, which was made spherical by ingesting a body larger than itself.

The senses of many hunters were incredibly refined. One early evening a raccoon stepped into the pond shallows to probe for frogs among plant roots. Though his feet were placed silently in the water, a flat wormlike leech, deep in the pond, raised his head in quest. He could not see the raccoon but was soon swimming rapidly toward him, and other leeches elsewhere in the water also headed for the animal. Before the raccoon left the pond, a dozen leeches had sucked blood from his legs.

If the hunters crippled their prey but did not catch it, other hunters would soon find and eat the victim. The young blue jay, recently drowned, was now being disposed of. Before the crows had been gone for a day, his body was surrounded by filaments of whitish creatures, visible only because of their great numbers. These water bacteria were eating the body, which created the stench of putrefaction. The body of the blue jay would pass through their digestive systems and would be returned to the pond as a clear, bland liquid that would help sustain other microorganisms in the water.

This was the epitaph to the hunter's victim and to the pond itself. In death there was life.

CHAPTER 7

the quarry

The quarry so well merged color, smell, and habit into the fabric of the pond that many of them disappeared from sight. They teemed through low patches of growth around the rim of the pond, resembling twigs, leaves, stalks, and bark. A chickadee flew into a bush and searched briefly, but found nothing and flew away. The katydids—green leaflike insects, whose evening song rasped over the pond—stood motionless and were not seen. Walking-stick insects looked like six-legged twigs; a large green luna moth blended into the leaves of a hickory tree; a white-lined sphinx moth dissolved perfectly into a background of hickory bark.

The pond was a maze of hiding places which absorbed the most unlikely creatures. A bass swam slowly through the water, and a crayfish saw him belatedly and catapulted backwards with such a rapid jerk that he seemed to dematerialize. The bass investigated a pall of muddy water where the crayfish had disappeared, but did not venture into it. When the mud cleared, the bass had gone and the only visible sign of the crayfish was his antenna and two beady eyes on stalks sticking out of the mud.

At the edge of the forest, a whippoorwill sat unnoticed on the ground. She had become a part of the earth as she warmed her three eggs with her long, hawk-like body. During days and nights owls, hawks, crows, and shrikes looked but did not see her. Above her, a mourning cloak butterfly became a part of maple bark, and a flycatcher, hovering nearby, did not see her.

The butterflies, despite their bright colors, instantly "disappeared" the moment they settled. Their caterpillars tricked the observing eye with red and yellow spots, black stripes, and hairy or tufted skins. The measuring worms, a type of caterpillar, fastened their rear ends to tree branches, reached up and dabbed some silk to the bark, then lowered themselves on it so that they stood out horizontally in perfect simulation of a twig.

The hidden life of the pond came into view slowly. The bullfrog's huge eyes looked out of the water among lily flowers, showing nothing of his body, which melted into the greenery around him. The varied hare's big brown eyes looked through long grasses, and his body was invisible. From a score of trees, birds looked over the edges of nests, waiting for a moment of violence as if pausing on the threshold of their lives. All around the pond, beady eyes watched, flanks moved in controlled breathing, antennae quested in the wind. The hunted life of the pond waited.

II

The sun shone on the empty husk of the beetle the robber fly had killed the day before. A tiny ant came out of the circular hole in the beetle's back. The tough horny skin of the husk was chitin, the almost universal shield of protection possessed by insects. Its strength and durability was shared by hunter and hunted. The

robber fly, though not possessing a skin of chitin, had a chitinous cutting tool the sharpness of which was regenerative, enabling him to cut through a score, or a hundred, beetle backs and still be effective.

In the larger beetles, the chitin was thick and heavy. In the ladybirds, it was light and delicate. Some chitin was so heavy that robber flies could not cut through it. Mosquitoes used chitinous cutters to pierce the skins of their prey and drink their blood. It was so durable that crickets, katydids, and cicadas, rubbing chitinous

119

organs together innumerable times a season, maintained a steady timbre of sound that buzzed and crackled—and sometimes roared because of the sheer numbers of singers—through all the warm days of the year.

Chitin was everywhere, in the empty skins of dragonfly larvae fastened to aquatic plants and in husks of cicada nymphs clinging to tree trunks. Chitin comprised the wing covers of flying beetles, the outer sheaths of praying mantises, the jaws of dragonflies, and the backs of bugs.

Chitin had evolved with the insects over aeons of time. The perfection of its evolution was marred by one flaw—its inability to stretch to accommodate growing bodies. The praying mantis had been killed by the yellow jacket wasp because of this, and every day many other creatures in the process of shucking off their old chitin skins were killed during the transition. All pond life had such flaws, existing with such predictability that they seemed to be a design of evolution.

But to escape the hunters, the quarry needed more than chitin. They used earth and wood tunnels and hid beneath stones or underwater. Some built shelters. The difflugia—one of the pond's rhizopods and a protoplasmic blob of jelly—built sheltering structures from grains of sand stuck together with a cement secreted by the animal. These shelters were pierced by holes through which difflugias could reach out for prey. In sandless areas, the difflugias fitted the empty shells of dead diatoms into nearly indestructible structures of protection.

Many of the tiny animals needed protection. Some

built their own, others produced it from within. One animal, the arcella, also a rhizopod, secreted from its body a brown shell of transparent membrane formed in countless minute hexagons. Another relative, clathrulina, fastened itself to a stone or a plant, then extended a long stem, which ended in a latticed sphere. Through the latticed openings the clathrulina extended parts of its body for food. Even plants could secrete shells, and the diatom shells were made in an indescribable profusion of shapes, colors, and sizes.

The wormlike larvae of caddis flies were scattered over the bottom of the pond, contained in some of the most effective portable shelters in the pond. Some of the larvae made their shelters from sand and cemented the grains together till they were almost completely encased in long tubes open at both ends. When sand was not available, the larvae used small sticks or cut up sunken leaves and cemented them together. These shelters gave protection against predators but not necessarily against other caddis fly larvae. One caddis worm having lost his shelter when its cemented leaf fragments broke up, swam to another shelter and bit the occupant in the tail. The bitten worm shot out of the tube, and the usurper slid into it to safety.

The caddis worms *seemed* to be safe; but nothing was secure at the pond. The water was now a fulgid green with the swarming life it held. The brightening sun exploded incandescence across the surface of the water, and a tiny fly came into the pond. She actually flew through the water, using her wings as though flying in air, and she alighted on a caddis worm case, inserted an egg in the larva's body, flew back to the surface, and disappeared into the air. The caddis worm was doomed to be host to the fly inside its body.

But most shelters did provide great protection. The greatest shelter builders were the wasps. Solitary potter wasps were now digging up drops of mud along the banks of the pond and molding them into spherical egg-containing shelters stuck to twigs and plants. Other wasps were fastening big mud shelters, with many entrances, to rock faces and tree branches. The paper wasps were chewing dry wood, mixing it with their sticky saliva, and producing a pulp that they were shaping into large and intricately built nests with dozens of individual cells for young wasps. The hornets built the largest shelters, also constructed of this dried wood pulp, some of them housing eight and nine thousand insects and containing several levels of cells, galleries, and tunnels.

The most enigmatic builders were the galls. Their shelters were swelling from branches, stems, and twigs everywhere. Tiny winged insects had laid eggs in goldenrods, in oak leaves, in the stems and leaves of dozens of plants and trees. When the eggs hatched into grubs, they secreted a substance that irritated the plant tissue. The plant cells went berserk and produced great swellings of tough-fibered tissue, some of them bulging out of leaves like a disease, others swelling stalks into graceful spheres. The gall grubs ate away at the plant or tree juices, protected from predators by the tree or plant housing them.

III

Blobs of foamlike material sprouted from the stems of many plants around the pond. Inside each mass of foam was a defenseless froghopper building the most ephemeral protection of any pond creature. The froghoppers had driven sturdy beaks into plants to extract a constant flow of food sap, and from their tails flowed a stream of tiny bubbles that pushed out into the mass surrounding them. The foam was less a shelter than a decoy. The froghoppers had to feed in open places along plant stems, and they could not outbreed those who hunted them. The bubble masses served to discourage or confuse the hunters for the short time that the froghoppers needed to escape.

A wasp hovered over a froghopper's mass of bubbles, and the froghopper withdrew his beak and waited in the cloudy soft light of his refuge for the swift rush of the hunter through the foam. His chances of escape were good because the wasp could not see his position. The wasp alighted on a stem above the bubbles. She would have one chance of making a kill. She flew suddenly toward the foam, and the froghopper heard the whine of her wings intensify sharply. As she burst into the foam, he shot down the stalk to the undergrowth, where he would find another feeding place and create a new shelter. The wasp crawled up the stalk and cleaned herself on a nearby leaf. Later, she dived into another blob of foam, and this time she emerged with a frog-hopper dying in her jaws. As she flew away with him, the foam was still oozing from his rear end. The pond froghoppers would later molt from their skins and fly off across the pond, no longer afraid of wasps.

IV

An earthworm waited just below ground level. Blind and deaf, the worm had his own sensory system. He felt the minute vibrations of a vole crossing the ground above him; and he waited patiently. He was full of digesting earth, sucked in through his mouth and pul-verized in his craw, the organic life in it withdrawn for sustenance and the masticated remains moved to his anus. He wanted to discharge this waste at the surface, but was in no hurry to show his pointed end above ground. He flexed his segmented body and thrust bris-tled hooks against the walls of his tunnel, pushing him-self toward the surface. The earthworm's senses were acute, but they missed registering the presence of a robin who was standing over him and listening intently, hearing the sliding noise of a body moving through the earth and calculating its exact position. Before the worm could push above ground, the robin's beak drove into the earth, seized him, and began pulling. Instantly, all the earthworm's body hairs fastened into the walls of the tunnel. The robin pulled, testing the worm's strength, and maintained the strain so the earthworm would tire. The worm slowly weakened. The hooks on his front end began slipping, and then half his body snapped free from the hole. The robin flew off, and the remaining half of the worm hastened deep into the soil to grow new organs and resume his subterranean life.

The ground surface bristled with peril for all the hunted who burrowed. A cricket lurked in some long thick grass looking out across a strip of sand at pondside. He remained in the shade, reluctant to venture over the sandy open ground that led to his burrow. Finally he started out in the hot sun, and instantly a humming, winged creature appeared over him with a zip of disturbed air. The cricket whirled around to face the yellow jacket wasp, dropping his abdomen to the ground and rearing up his forelegs. The two insects remained facing each other for moments. Then the wasp whisked away, and the cricket scuttled across the open ground to his burrow.

Bluff was a common quality in the hunted, but many also had real weapons to ward off hunters. A sow bug, pursued by a pair of tiger beetles, drove them off with a jet of fluid from his rear end. A monarch butterfly attacked a bumblebee and drove him away from some nectar-rich flowers; a bumblebee drove a hummingbird from the same flowers later in the day. But the quarry seemed never quite equal in guile to the hunter.

A red-legged locust clattered into the air from the sandy peninsula and headed across the pond. He was seen by a flesh fly, a gray-colored insect who was a tiny fraction of the locust in size. The locust never knew that she had landed on his back and laid a dozen live mag-

gots there. He flew south to the marsh, and the flesh fly
returned to the pond. No eyes saw the gradual slowing
of the locust's life force as the maggots ate into his non-
vital regions. Eventually, as the maggots grew to maturity, he would become immobile.

The most confident hunter might find his assurance
upset by the unexpected strength of his prey. A snake
one morning swallowed a large cicada nymph that had
just dug out of the ground and was heading for the
nearest tree to undergo transformation into a flying
adult. The snake slid along the pond's south bank, but
soon showed signs of distress and began turning in one
direction and then another. The nymph, whose blindly
instinctive movement toward his place of fruition had
been delayed, was not dead or injured. He was, instead,
faced by an irksome barrier. For years he had dug through
the recalcitrant ground, and now he was still trying to
dig. He drove his powerful forelegs into the snake and
lacerated tissue and muscle. The convulsive muscular
contractions of the tunnel in which he was trapped interrupted his digging, and he was soon squeezed into a
moist gallery. As the snake continued his agitated passage, his stomach juices finally killed the stubborn cicada
nymph.

V

The inexhaustible numbers of the quarry fed the
whole pond. If their torn, punctured, poisoned, drugged
bodies did not feed the initial killer, they would surely
feed the secondary hunters. Thousands of tiny crusta-

ceans roamed the bottom of the pond in search of the remnants of bodies torn to pieces in the hunt. On land, millions of ants scavenged endlessly for dead and dying butterflies, cicadas, flies, and mantises.

A late homing ant found a dying dragonfly early one evening and examined him for a moment before racing to a distant ant mound near the pond. A group of ants soon emerged and went back following his tracks. When they reached the dragonfly, they began dismembering him. Nothing was wasted. The legs, wings, head, and thorax were separated, and there were precisely enough ants to carry the pieces to the mound. By dusk, the dragonfly was underground. But the ants left behind a puzzle. To reach the dragonfly, they had followed the trail of scent exuded by the original ant. But how had the size of the victim been communicated?

As the ants dragged the pieces of dragonfly home, they had passed near the headless blue jay chicken, and a burying beetle had clattered over them on his way to the body. His life, too, depended on finding dead bodies. With his red-spotted wing covers stiffly outstretched, he came noisily through the trees and dropped down near the blue jay. He examined the bird carefully and ran around and over it several times. Then he scuttled some distance from the bird and began digging a hole in soft ground. His powerful head and limbs worked efficiently, but before he disappeared from sight in the hole, he ran back to the bird's body, threw himself on his back, and pushed rapidly out of sight underneath. The body trembled and moved slightly. The action was repeated till the beetle, still on his back, appeared on the other side of the bird. He had been lifting the bird toward the hole by using his feet as levers, and he scuttled around to the front of the bird to repeat the process.

A second beetle clattered out of the darkness and joined the first with the same urgent purpose. She examined the hole, ran back to the moving body, rolled over on her back, and pushed underneath it. With twelve legs thrusting against it, the bird's body moved ahead faster.

By midnight, it was poised on the edge of the shallow hole. The beetles rolled it in and were then ready to begin the most difficult part of the burial. Each disappeared under the body, pushed out soil from underneath and lowered the body into the ground with visible speed. They kicked the soil they had dug over the body, completely burying it. Soon, the bulging of the newly dug earth was the only indication that the creatures were working there. Underground, the beetles were stripping off all the feathers and packing them away in the loosened soil. The female was ready to lay her eggs, thus beginning the next step in the use of the blue jay chicken.

At rare moments, hunter and hunted came together in a flash of action combining the supreme effort of each. The varied hare, keeping his endless watch on sky and undergrowth, one day saw the red-tailed hawk suddenly loom from the opposite direction. For a moment, both creatures seemed caught in time. The hawk's talons were swinging forward and his powerful neck was arching to bring them together at the body of the hare.

The hare was tensed into a bulge of muscles and eyes as his whole being concentrated on the kick that would jerk him away from the attack. The muscles worked just when the hawk was no longer able to alter his stoop and he struck ground heavily. When he righted himself, the hare had gone.

CHAPTER 8

the vulnerable seeds

The first epic of the year at the pond had been the re-
covery from winter. A great new struggle was now in prog-
ress. The pond renewed itself every year with a prodigal
loss of life. The blue jay chickens were hunted to their
deaths. The pond's six adult yellowthroat warblers were
killed by owls, and their three nestfuls of chickens died of
starvation in the sedge and marsh grasses. Every day,
hawks, minks, and weasels destroyed a dozen red-winged
blackbirds' nests. In thirty days, more than twenty thousand mice had been born in burrows among logs and
roots around the pond, and only a thousand survived at
the end of that period. Ten nests of robins had yielded
five surviving young birds. In reaching for the future, the
pond creatures were so often frustrated that survival there
was clearly an exception of life.

One hot evening, thousands of gnats massed over the
polished surface of the water. They flickered through the
air, rising and falling as they copulated. The fertilized

females bombarded the water with innumerable white spherical eggs, which fell through the water skin to the bottom of the pond. A newt swam into the rain of eggs and began eating them and was soon joined by other newts. The old muskrat, swimming home to his mate and four youngsters, scattered a group of elvers, or young eels, who had risen from the mud and were tracking back and forth, engulfing thousands of eggs through their wide-open mouths.

The gnats were being decimated by dragonflies above the water, and before they disappeared into sleeping places in the trees, about a quarter of them had been eaten or had fallen into the water. Most of the eggs had been eaten. Five hundred were united by chance on the water skin near some lily pads and drifted into a mass of aquatic plants. They would soon sink and might survive among the tangled roots below. The gnats, not possessing tooth or claw, speed or agility, survived by reproducing in vast numbers.

In contrast, a small, gnatlike fly settled on the surface the next morning and laid a cluster of eggs, which she lowered into the water on the end of a fine line. She ended the line with a flat cap, which caught on the surface skin. Then the submerged eggs drifted off.

Some creatures reproduced in billions, like the diatoms and desmids welling up in the water like a mysterious animation of plant life from nowhere, and others produced a few score progeny, which they might have to guard to ensure survival. Male sunfish hovered over their mates' eggs in pebbly nests on the bottom of the pond.

Their concern for the safety of their eggs was matched by other, simpler creatures. Forty young leeches clung to their mother by suckers in their tails, without which they would have small chance of surviving. A female sunfish, her egg-laying finished, discovered that she could dart forward and wrench away odd youngsters without risking the mother's life-sapping bite. But some young leeches would survive.

Some creatures fought for the survival of their children. Female earwigs, sheeny brown insects with large posterior pincers, brooded over their eggs or their newly hatched youngsters. Underground, another mother was showing unexpected devotion. A female mole cricket—one of a hundred or so who lived around the pond—had built an underground nest and laid about three hundred and fifty eggs in it. She was crouched over them, black and shiny, her clawed feet and heavily armored back set in a protective stance, waiting expectantly for the scuffling, scraping sound at the entrance of the burrow which would signify the arrival of a ground beetle whose taste for mole cricket eggs transcended his natural repugnance to entering a strange burrow.

Later, the mole cricket would have to fight off another marauder, the male mole cricket, who had a taste for his own eggs. There was little chance of her rearing her youngsters this year. Crickets were already overly abundant at the pond, and pond resources would soon be stretched to feed them all. The female mole cricket would be long hunting, and eventually the male would eat the eggs while she was absent from the burrow.

The cool earth also harbored the bumblebees, who, later in the season, would also regulate their numbers. Earlier in the year, the queen bumblebee who had slept in the granite rock found a field mouse's burrow near the edge of the water. It was deep and dry, and she flew far into it before settling and scuttling to its farthest end. During ensuing days, while carrying dry grasses into the burrow to make a nest, she met an inquisitive vole at the burrow entrance or inside. Once, she flew in the entrance while the vole was deep in the nest and the two creatures faced each other, the vole a humped indistinct mass of fur, the bee showing dimly yellow and black silhouetted against gray surface light. The vole had to pass the bee to escape but was intimidated by her loud buzzing and confident aspect. His expectation of finding honey in the burrow, as he had in the past, was unrealized, and now he wanted to get back to the bright surface. In a sudden excess of courage, he rushed the bee and in a flurry passed her without being stung.

The bee spent the early season building a wax honey pot larger than herself from materials secreted between her body segments. In warming, blossom-starred days, she filled the pot with nectar and pollen and then began making wax egg cells. In each she laid one egg, covering it with pollen and sealing the cell with wax.

Then, looking like a small brooding bird, she hunched herself over the egg cells. Except for odd expeditions for food, she remained there till the eggs hatched and the grubs began spinning their cocoons. Then she waited till they emerged from their pupae. They were miniature bumblebees, perfect replicas of their mother and long-dead father. They staggered over to the big honey pot, drank deeply, then scuttled back to their mother.

II

The nights were filled with wings. The bats flew high and silent, appearing erratically against the light of the moon. The polyphemus moths, almost as big as bats, flew low. Like the May flies, they were too vulnerable to survive as adults. They neither hunted nor ate, and their brief moonlit lives were for mating and oblivion. Within a few days of emerging from chrysalis cases scattered through the forest, the females were laying egg clusters under leaves. The big moths were dying long before the eggs hatched. A dozen fell near oaks on the north bank and were eaten by some raccoons. Many crawled into tree holes or forks in branches or hid under leaves, and their bodies passed insensibly from the pond. One moth was harried from the oaks by a sparrow but made little effort to escape. He jerked downward and struck the water and drowned without a struggle. The sparrow looked down from an overhanging branch. The great outstretched, boldly eyed wings looked back, and the sparrow flew away.

The apathy of the polyphemus moths was not shared by their youngsters. Even as the adult moths died, clusters of disc-shaped eggs began hatching and redheaded caterpillars gnawed clear to expose beady eyes and long yellowish bodies. They ate immediately with intensity and speed from the first leaves within reach—chestnut, maple, birch, beech, elm, oak, sycamore. Some ate fifty thousand times their own weight in the first two days of life. Their green droppings pattered down in a steady rain, and their bristly bodies, ornamented with touches of silver and red, humped on endlessly to new leaf pastures. Under them, one night, young burying beetles, who had pupated after eating the jay chicken underground, dug clear to the surface and flew into the silent forest.

III

In seeking the best protection for their youngsters, the pond creatures revealed the extent of their sagacity and development. A pair of otters in the marsh had dug a long tunnel that branched and rebranched and had submarine and terrestrial exits. A weasel had a fur-lined nest under a heavy stone. A pair of oven birds at the swamp had a tiny spherical nest with a matted grass flap covering its entrance.

However elaborate or secure protection might seem,

135

there were always some creatures who had developed methods of nullifying its purpose. During the previous year, scores of horntails and sawflies, which looked like large handsome wasps without nipped-in waists, had buried their eggs in dead branches throughout the pond area. The eggs had hatched into large white grubs, which bored deeply into the dry wood, leaving no outward sign of their presence. But another pond creature knew they were there.

An unusually shaped, yellow-faced black insect flew slowly over the pond. When she landed on a pond weed her wings stood erect from her slight shoulders and her thorax arched in a complete semicircle, with its tip pointing to the ground. Her antenna was long and yellow, and her legs were equally long, two of them supporting the front of her body and four trailing to the rear. From the tip of her downcurving abdomen projected two hollow tubes that described a complete semicircle and came down past the insect's abdomen. This was an ichneumon fly, and she was looking for sawfly and horntail grubs.

She flew into the green leaf dome of an elm toward a dead branch that jutted over the pond. She settled on it and listened. She could hear the vibrations of a moving grub deep in the wood, and she calculated his position from the sounds and moved slightly twice before dropping her curved black tubes to the wood. With imperceptible speed she pulsed the ovipositor into the dry wood. The drilling continued for a quarter of a morning till the ovipositor broke into the chamber of a pigeon horntail larva. It pierced the larva's body, and an egg came down the tube.

The ichneumon withdrew her ovipositor and flew off, leaving her grub to eat the horntail grub. The boldness of her conception invited danger. Elsewhere on the dry branch, an ichneumon was dying. She had drilled so deeply that her abdomen touched wood but her ovipositor was still short of the grub's chamber, and she could neither drill farther nor withdraw. Nearby, another fly's ovipositor fluttered in the wind, tipped with the wrenched flesh of a female who had pulled free but who would later die.

IV

The pond sent its creatures to unsuspected summits of ingenuity and energy in this season. The communal and hunting wasps reproduced with a semblance of intelligence which seemed unsurpassable but which was matched by the creatures who preyed on them. A digger wasp bumped down on the sand on the peninsula. She dug frantically with her forelegs till the tunnel was ten times the length of her own lean body. Then she hummed across the pond, her eyes glistening for the sight of a grasshopper. She saw one and paralyzed it with a sting. Though the insect was nearly as big as herself, she straddled it and strained into the air. She bumped down at her burrow, dragged the grasshopper inside, and laid an egg on it. Immediately, she burst from the burrow and was off hunting another grasshopper.

All the pond wasps seemed to share the digger's urgency. The yellow mud daubers finished fastening upright mud cylinders under a rock ledge. In each chamber was a paralyzed or dead spider with one wasp egg. The mud daubers disappeared, but not before they had been observed. Soon after they had finished sealing up the last cylinders, a slender-waisted blue wasp drank at a lily leaf. She drank till her body was distended and the last droplet of water glistened in her jaws. She flew up to the cylinders, and soaked a clay wall and returned to the pond for more water and soaked again, and chewed and soaked till she broke into the chamber. She hurriedly pulled at the spider till she found the yellow mud dauber's egg. She ate it, pushed the spider back into place, and laid her own egg on the body. Then she sealed up the hole and disappeared across the pond.

The creatures of the pond were driven on by a force of life that could make them frantic. One wasp who was dragging a spider to her burrow was almost comically nimble. She tugged at the body with such haste that she jammed it momentarily in the entrance to the burrow. That night, after laying only one egg, she clung sleeplessly under a leaf. Her abdomen twitched, then bulged as life twisted inside her. She clung tighter to the leaf, but soon after dawn she dropped dead, wings rigid, legs curled. Her egg-laying had been delayed because of a shortage of spiders, and in the heat of her desperate hunting she had accelerated the development of the eggs in her body. The young carnivorous grubs had begun hatching in the night and immediately ate into their mother's body. She was dead, and soon they would be too.

In the blazing sun of the next day another wasp was engrossed with a different kind of problem. Unlike some other wasps, the female black wasp could not fly with her victim and was dragging it laboriously along the edge of the pond. She tugged and strained, falling backward into burrows, snagging the paralyzed spider in grasses, and once, aided by a puff of wind, actually flying backward while the spider rolled along in a bumping spurt. She blundered into a spider's web and pulled free, without seeing the tiny spider watching from a slit in the ground. The sun moved across the sky, and the black menacing shape of a robber fly appeared suddenly against it. The wasp darted up, and the robber fly flicked out of sight. The wasp dragged the spider four thousand times her own length beside the pond, till, leaving a drag mark across the yellow sand, she disappeared into her burrow. It was hot and the pond whispered. The wasp reappeared and flew across the water in search of more spiders.

On the far side of the pond, another wasp was rolling a spider into the water. She climbed on his body, gripped it with her legs, and took wing. The spider surged away from the shore, and an underwater frog kicked up to examine the spectacle. He might have grabbed the spider, but his attention was distracted by the humming creature flying just above it. The wasp reached the far shore, rolled the spider up a sandy incline, and pulled him into an earth burrow.

139

All the wasps were industrious and ingenious, and the cicada killer wasps were among the most highly developed. A female cicada killer had dug a deep burrow on the pond's north bank, and now she was hunting for cicadas. She followed the sound of a big green cicada through high sparse branches, and the cicada, sensing danger, abruptly stopped singing. The wasp drifted through the leaves, the sun flashing on her great broad-veined wings as she deliberately dropped down on her victim. The powerful insect sprang from the branch, carrying the wasp with him, but he had been stung and his nervous system was paralyzed. The two insects fell straight to the ground; yet the fall did not break the wasp's grip. She had her victim, but obviously she could not fly with him. She was separated from her burrow by the widest part of the pond. All around were thick bushes. But the cicada killer had no need to pause: she dragged the cicada to the nearest tree, mounted a root, and began the vertical climb up the trunk. The pond became silent, shadows moved through the trees, and the forest breathed inaudibly as the wasp edged up in her herculean climb. The greenish flowers of the bunchberries became small and indistinct as she climbed and came into sight of the burnished pond and the blossoms of pickerelweeds and arrowheads that showed through the leaves. A kingfisher drew a blue line across the pond, and the wasp ascended into the hushed canopy of leaves, where she was surrounded by heedless caterpillars. A scarlet tanager flashed by. The entire pond was now in view, and though the wasp did not look for it, the entrance of her burrow was faintly visible on the far side of the water, a tiny shadow in the sun.

She paused at last, but not for rest. She twisted the cicada on to a branch, climbed on top of him, and with her legs wrapped around him tightly, launched herself into space. At first, it seemed she had miscalculated. Her initial drop sent her plummeting toward the water. But she leveled out slowly and, though still falling, was now halfway across the pond and the strident hum of her wings sent dragonflies flicking from her path. She caught an updraft from the warm water and collided violently with a sand patch near her burrow. She did not pause but hauled the cicada toward it.

The wasp's massive build and powerful flight had evolved through the ages to enable her to procure meat for her youngsters. But another creature, only a fraction of her size and strength, had been evolving with her. The wasp left the cicada for a moment at the burrow entrance and went inside, then reappeared and began pulling the insect toward the hole. As the cicada filled the burrow entrance, a parasitic fly, who had watched the arrival of wasp and victim, flew from a twig, alighted on the cicada and planted one egg. The fly was one of hundreds of creatures at the pond who had learned to let others do their work for them.

V

The male mole cricket finally ate his mate's eggs while she was hunting. In the pond, a female great diving beetle was trying to ensure that her eggs did not meet the same fate. She was holding her mate tightly while carefully laying and cementing a mass of eggs to the hard horny shell of his back. Perforce, he would guard the family, but only because he could not reach the eggs. Like all of his kind, he had an implacable appetite.

Another male diving beetle paused on a sunlit bar of sand nearby. The eggs on his back were hatching, and the young, black-eyed, white-bodied beetles were straining out of their egg cases. The first one free turned to the nearest youngster, gripped him between biting pincers, and began eating him. The young beetles oozed forth, fighting and struggling and spreading out into the water as a living example of the invincible instinct to hunt and kill.

Nearby, toward the crystal surface that was so vividly illuminated by the sun, a colony of hydras writhed their stemlike limbs from the bur reeds. They were reproducing by budding, and the youngsters swelled under the adults' skins and burst out as miniature hydras. The young hydras remained attached to their parents and hunted as relentlessly as the adults. A water flea moved toward the hydras and was seized by an adult, but the catch was contested by a youngster growing from the adult. The two animals wrenched and ripped at the prey till the youngster took the flea from his parent and swallowed it.

The young hydra was too young to have his own digestive system, and the flea passed into the body of the parent for digestion. Later, the young hydra would break free and begin independent life.

Some of the four hundred odd surviving crayfish in the pond were laying eggs. A female reared up on a submerged rock so that her head was high and her tail low. She cleaned her abdomen, then turned on her back. From her tail she passed a milky white material which, with her claws, she spread evenly across her abdomen. Then she curved her tail, exuded about five hundred eggs onto the milky mass, where they stuck. The crayfish righted herself and flipped off through the pond.

Clams opened their shells and extended one long white foot, dug it into the mud, and pulled themselves forward. The half-open shells revealed hundreds of eggs packed tightly together. When the eggs hatched into tiny shelled creatures, they would leave their parents and fasten themselves by hooks to passing fish and so travel upstream to other ponds and downstream to marsh and lake. Wherever they ended up, they would drop into the mud and begin growing to adulthood.

Pond lilies were speckled with leaf beetles, metallic green, bronze, and purple creatures who cut small holes in leaf centers. One beetle inserted her abdomen in the hole and fastened a circle of eggs to the underside of the floating leaf. Within a few days the leaf beetle larvae would hatch, and then, instead of heading for the surface, they would wriggle deeper into the water, even though they could only survive by breathing air. They would fasten themselves to plant stems, dig holes in them, and put their heads inside. Tiny globules of air contained in the plant stems would bubble to the surface, and the grubs would settle down to eating plant tissue and breathing oxygen produced by the plants. Later, they would spin watertight cocoons, each one filled with air

143

from the plant stems. After they had pupated into adult beetles in their submarine cases, they would cut through the walls, and a bubble of air, adhering to the silken hairs of their bodies, would carry them to the surface. If for any reason their emergence into the sunlight was delayed, they could breathe oxygen from the bubble.

The leaf beetle finished laying her eggs and flew away, and an eye watched her from a patch of grass. A nest of young cottontails was hidden there, partially covered by a layer of rabbit fur. When the female cottontail returned later, she pushed aside the fur and settled to allow the youngsters to suckle.

VI

A leopard frog floated under a leaf in the southern shallows, motionless, with the patience of his kind. He saw the water dimpling ahead of him, and at each dimple an egg came into the pond. Above, a score of dragonflies laid eggs, dipping their abdomens precisely into the water. Many of the eggs reached the bottom of the pond, and many were eaten. Near the leaf above the frog, the water parted and a small dragonfly appeared. She swam past the frog, who, because he had eaten recently, ignored her. She clung to the stem of a plant near the bottom of the pond, sustained by a bubble of air that glistened on her body. But the air was dissolving, and she had to move quickly to finish her task or be drowned. She fastened a dozen eggs to the plant stem and headed for the surface.

She left the water as other insects entered it. Wispy damsel flies, who looked like small dragonflies, came into the pond in pairs, the males gripping the females' heads with their abdomen pincers, and drove down through the water to alight on stems of aquatic plants. After the females had laid eggs on the stems, the insects swam back to the surface, where the males broke through the water skin and paused for a moment, their transparent wings spread in the sun. Then they lunged forward against the resistance of the water skin that held their mates underwater, and suddenly the insects slipped free and disappeared across the pond.

Insects whose lives began in the water had to lay their eggs there, but the rate of survival in a pond swarming with egg eaters was low. Some creatures even laid their eggs on the eggs of other creatures, and if the intruders hatched first, their larvae would eat the host eggs. Tiny flies were entering the pond now and paddling down in search of sunken dragonfly eggs, and one type of ichneumon, similar to the wood-boring creatures, came flying through the water on a similar search. Soon after an aquatic back swimmer had finished injecting some eggs into the stem of a pond plant, a fairy fly laid one egg inside each back-swimmer egg.

The egg layers tried to outwit the egg hunters. Some produced overwhelming numbers of eggs. The stick insects, spread widely through the forest north of the pond, pumped out half a million eggs in one afternoon, creating an almost continuous pattering on dry leaves below. Chalcid flies laid single eggs, each able to produce more than two thousand young flies. The midge larvae, now hatching from eggs, were victims of the same urgency of

145

life which earlier had set the young wasp grubs to eating their mother. The egg-producing glands of the midges developed faster than the rest of their bodies, and before they could escape from the pond as adults, their bodies had burst open, flooding another generation of larvae into the pond.

VII

The pond was a study in contrasting methods of survival. Young foxes, hawks, and owls were produced in small numbers, and most of them survived, at least initially. A denful of fox cubs rolled and played in the sun on the northern slopes of the pond. Two young red-tailed hawks stood gawkily erect in an untidy nest atop an elm west of the pond. Two young great horned owls, blotched brown, had long since left a nest stinking with the vomited, pelletized remains of feathers, skulls, and skeletons of rabbits, mice, and birds. When they left, their parents methodically tore the nest to pieces. One night, a young horned owl settled in a low bush near the pond and looked directly at the fur-lined nest of the varying hare. Being inexperienced, he did not see the nest; the female hare, rigid with fear and concern, watched from the edge of the forest as he flew away.

The foxes, hawks, and owls were masters of their environment. But all living things sought to shape environment to their own ends, and nowhere was this better personified than among the plants around the pond, whose flowers were to be seen now. The rich-flowering purple thistles would shrivel and dry and later send out

146

thousands of seeds suspended by air-borne carriers of gossamer. Wild oats, reaching up thickly along the stream entering the pond, would produce seeds that would be dropped or blown into the thick grass. When the first rains hit the seeds, an elbowed arm on them would straighten out and likely thrust the seeds deep enough into the vegetation to reach the earth.

Many of the plants at the side of the stream would drop their seeds into the water, and the seeds would

drift into the pond on built-in floats that might take them to a landing and—remote chance—germination. The milkweeds, growing so thickly on the peninsula of the pond, would release flying seeds like the thistles. The milkweeds could measure the force of the wind. Each flying seed remained attached to the pod by an adhesive spot that could only be loosened by a puff of wind. The seed pods would roll back their cases gradually so that the seeds, arranged in tight spirals, would be loosened separately, each one drying perfectly, then soaring away.

The fruit pods of the tea plants would explode seeds into space. Ripe jewelweed seed pods, which resembled small beans, would burst from the plant at the touch of a fox's foot. As the pods flew through the air, a series of uncoiling springs inside them would scatter the pod seeds over a wide area.

The wild geraniums, destined to decorate the woods with pink flowers in late summer, would turn their blossoms into compact five-sided pods. As the seeds ripened, the pod would be pushed upward, becoming purple and ripely polished, and then would explode, sending the seeds in many directions.

Some plants used dissimulation to ensure survival. The woodland violets, now sprouting thickly, were easily crushed by the feet of passing animals. If their fine-stemmed flowers were destroyed, they would switch their growth force to other, petalless flowers on stalks at ground level. The seeds of these flowers would ripen into oval capsules tinged with purple. As the fruit matured, the stems holding the capsules to the plants would stiffen and lengthen. The seed capsules would split into three segments and contract onto the hard seeds, and eventually the seeds would be pipped forcefully out of their containers.

The principle of pitting tension against tension was most successfully used by the wood sorrels, whose three-leafed clumps were blanketing great areas of the northern forest, along with trilliums and bunchberries. The

148

sorrels would later push up small clusters of five-petaled yellow and blue flowers, which would turn into wet, green seed capsules. These capsules would be triple-layered, the outside layer being hard, the center layer soft, and the inner layer normally thin-skinned like other seeds. At seed time, the plants would send water up their stems, and this would expand the center layer. At the same time, the outside layer would be contracting quickly. Suddenly, the capsules would turn themselves inside out and fling their seeds far into the forest.

In these, and other ways, the pond plants revealed close resemblances to animal life. They overcame many of the same problems, and their drive to reproduce was just as strong. Throughout the damp southern forest, slime molds were growing into prominence. Each was a group of bacteria-hunting cells, neither animal nor plant. Their slimy masses favored dark places and seemed to have no central intelligence but could move toward a common objective. The molds were crawling slowly from the shade of hollow stumps and from behind damp patches of bark toward warm, sunny places. The front cells deposited slime on leaves, sticks, and stones; and the following cells, bunched thickly behind, rolled over them. Later, the molds would stop their traveling. The front cells would fasten themselves to any dry, warm place they came upon, and others would pile on top of them, building up a slender column of cell life. At a certain height, the cells would suddenly begin dividing and redividing to form a large sphere of cells at the top of the column. The cells would ripen in the sun

149

and dry, and then the sphere would burst, sending countless millions of spores floating through the forest and over the pond in search of places to bring new colonies of slime mold to life.

VIII

From a hole high in a partially hollow oak tree, a bright pair of eyes peered for a second; then a wood duck emerged quickly and flew to the pond. The hole gaped blackly behind her. A downy young duckling appeared and looked at a giant's distance to the ground. The wood duck circled the pond; her mate quacked in a nearby tree. Another duckling appeared at the tree hole and jostled the first. Then, as though propelled by a spring, the first duckling flung himself into space, tumbling through the silent leaves, plummeting for agonizing moments. The second duckling followed while the first was in mid-air. The first duckling bounced on a log, the second crashed into a bush. Others were spewing out of the hole, jumping hastily now. The great fall hurt none of them. Before the sun had gone an eighth of the distance across the sky, the ducklings were following their mother across the pond toward swamp reed beds.

The hatching of the ducklings coincided with the arrival at the pond of a snapping turtle. The turtle swam slowly and awkwardly up the outlet stream of the pond. He was larger than any other pond turtle, being about as big as the hen duck herself, with a heavy carapace thickly studded with slimy green algae, and undersides colored pale yellow. His long tail, heavy armor-studded neck, and long claws jutting from his powerful legs gave him an appearance of power and purpose. He surfaced near a cluster of lily pads, his star-pupiled eyes and blunt snout barely protruding from the water. The duck and her brood, passing nearby, did not see him.

They did not see the circle of water moving from where he submerged or see his flat green shape gliding

under them; they did not see the blank eyes looking up, or the shape looming larger. But with a suddenly cackling, peeping spurt of terror, they did see a duckling jerked underwater. They scattered into the reeds, and the duck took wing and circled the pond, quacking agitatedly. For the turtle, the duckling was only a mouthful, and he searched the reeds for others, spotting the rounded downy bodies of three youngsters huddled,

151

trapped almost, among some closely packed spears of aquatic grasses. They were well hidden from any but a submarine hunter. The turtle ate them quickly.

The duck, who had no experience with turtles, knew now that something dangerous was underwater, and she landed on the pond again only after the most circumspect delay. She called for her surviving ducklings, and the three of them quickly jerked out of hiding and fell in line behind her. But the turtle was still watching, a motionless green mark on the bottom of the pond, and as the four shapes swam against the spangled sun, he rose swiftly and took a duckling. The duck had led her family to the spilling outlet, and though she shot ahead when the duckling went under, she did not fly. Her ducklings struggled behind, and the three surviving birds bobbed down turbulent water to the marsh. The duck now knew the dangers of water and would keep her ducklings in the shallows around sedges and reeds, where submarine hunters could not take the rest of her family.

The trees were filled with young birds who were as vulnerable as the ducklings. At dusk, the woods between the pond and the marsh tingled with the mellifluous runs and trills of the hermit thrush. It was a song of survival and of re-creation. Death did not break the promise of endless generations of young creatures springing into life at the pond.

CHAPTER 9

the center of the universe

The pond was now a vision of brilliance, a luminous pulsing mixture of greens, reds, blues, grays, yellows, browns, purples, pinks, the colors of the earth. Lilies garlanded the water with spiked white flowers, and water hyacinths clustered delicate yellow blossoms around them. This florescent time revealed the richness of life at the pond. Bee balms spattered red flowers around the cricket colony on the sandy peninsula and along the banks of the stream. Jewelweeds rose from their own thick green mats of growth near pond and stream. Clumps of thistles on the sandy peninsula, spiked and green and sumptuous, nodded purple flowers in sleek anticipation of forthcoming fruition.

The trees swept silently around the pond above and below its shoreline, and their inverted pictures wavered and shook in the water. In mornings and evenings, brilliant lights flowed through the odd spaces between them, appearing in the pond as patterned shafts that shimmered as wavelets fled overhead.

In the far shorelines, among cattail leaves and bur reeds, the luxuriance and depth of foliage of the pond melted into the forest and obliterated the division between land and water. The unknowing eye saw the stillness and missed the vortex of life within, missed the falling rain of honeydew from bulbous green aphids, missed the marks of crouching moths, the hiding mosquitoes and flies, the brooding birds, the chomping caterpillars, the climbing snakes, and the swift flight of the sparrow hawk, who in sunlit spaces was a spasm of motion.

Only the red-tailed hawk was clearly seen as he looked at the flashing reflection of the pond and at the northern woods of spruce, hemlock, elm, beech, birch, and pine, a variegated pattern of greens patched by rocks in treeless places. The hawk watched birds flying through the southern forest of birches, beeches, and maples and saw snakes gliding over dry ground into long grass.

His small shadow stopped for a moment at the pond's western end, where the stream spilled into the pond near the sandy peninsula, creating an egg-shaped swamp thickly grown with sedge, rush, cattail, and pussy willow. The hawk looked down on red-winged blackbirds and faintly heard their harsh chattering as they flew rapidly between swinging sedges. He saw the blossoms of the pickerelweed and arrowhead and saw catfish crawling over shallows at the marsh edge of the pond, but he did not see them laying yellow eggs on stones.

He saw submarine frogs swimming erratically underwater and marsh ferns spread along the edge of the swamp. Water snakes basked in the sun, and a muskrat swam across the pond. The hawk heard the deep rich rumble of the bullfrog calling vainly from a lily pad for a mate, and he was used to seeing wood ducks flying quickly through the northern forest and disappearing into tree trunks, but this time he was not watching when the ducklings emerged. He watched a female grouse, followed by a dozen chickens, bobbing through the thick grass near the swamp, but he ignored them and soared away slowly to the south, where the wide shining scar of the marsh burned the sun back into his eyes.

II

A screech owl waited one night in a huge elm overhanging the pond. Few of his prey ever saw his tufted ears and short tailless body. His yellow eyes watched the pond, which swarmed with nocturnal life. Bat wings flitted silently against the hovering moon; fireflies burned and died in thick masses of bushes and branches; huge moths came sweeping through the darkness and seemed to put out firefly lights with the wind of their wings.

The night was an active time, even by starlight, when sensitive eyes perceived the configuration of the pond as though it were day. The screech owl waited, watching mice appear from their burrows and listening for scuttling movements in the leaves. Thousands of their progeny were hidden underground ready to feed pond and owl. For a moment the owl looked above the trees to the stars, where he could see the faint marks of clouds and where soon, he knew, the moon would fully illuminate the rushing life in the air. Nighthawks crisscrossed the sky, beeping loudly. A torrential rushing sound made the owl crouch, but the nighthawk, having dived from a great height, suddenly swooped up with deeply resonant humming, which faded to a murmur. The owl launched himself across the pond.

His view of the night life of the pond changed sharply once he was in the air. He found himself among multifarious moths. Hawk-shaped sphinx moths sped past him, and a polyphemus moth flapped overhead. Before the owl reached the far shore, he had seen ios, imperials, great lunas, and white acrea moths.

The bats flew mad courses through the moths, hitting them with soft slaps and biting into their bodies with splinter-sharp teeth. The air was filled with thin, high bat cries, which were audible only to ears that received the incredibly high pitch. The calls bounced off all flying insects, and the reflected sounds informed the bats of distances, directions, and speeds. Their crazy, zigzagging flight branched from mosquito to moth in blind destruction of flying life. The bats lived in an almost completely dark world of echo.

The owl ignored them and landed in black southern trees. The forest whispered with movement. A whippoorwill called and was answered from the marsh, and a third call faintly sounded in the east. A flying squirrel drifted between trees, floating on webs of skin stretched between her outflung legs. A fox trod on a twig and snapped all life into frozen expectancy. Three meadow mice stood rigid by a hickory root. A muskrat in the swamp whacked his flat tail on the water and dived. The motionless owl suddenly perceived the three meadow mice; his feathers shrunk against his body, and soundlessly he floated down. A strangled squeak was lost in the dark.

The life of the night sent inexplicable noises out of the reeds, and frogs pattered into the water at the suggestion of an alarm. The bullfrog floated alone and watched a rim of incandescence appearing over the trees, and the pond was suffused with moonlight. A terrible scream hung in the distance as a marsh mink, long engaged in a stalk, finally ended it in the glow of the moon by pulling down a heron in shallow water.

The moon penetrated the forest, and an unexpected burst of daytime bird song rippled through the trees as sleepy cardinals sang on being roused by the light shining in their eyes. A sparrow woke and sang. Chips of song scattered into the white light, and redwings called restlessly. The screech owl floated through high branches to the marsh.

The eerie expectancy of night had vanished in this strange, daylike brightness. Insects whined and rasped, and the sounds seesawed against singing crickets who clicked and ticked and wheeped and groaned. A snowy cricket sang under an elm, and others of his kind joined him, creating an uneven sound. But the crickets unified their singing till it had a steady, pulsing beat.

A narrow-winged cricket sang a mournful song among

the hemlocks. Nearby, a dozen striped tree crickets whirred incessantly. Deep in ground tunnels, crafty mole crickets sang, securely hidden, and their growling calls came from everywhere in the earth with the ventriloquism of all cricket song, so that now a call came from under some trilliums, now from some elm roots, now by the crinkling stream.

In alder thickets on the pond's north shore, winking firefly lights were visual, rather than audible, communications. A male firefly flew over the pond, watching a light flashing on an alder leaf. His whole abdomen glowed, and the light on the leaf winked responsively. He flew to the leaf and ignored other lights now appearing around it and landed beside the female whose light had beckoned him.

The moon was now enormously white, creating filigreed light patterns that shone through the overhead leaf cover. As it moved up into the dark blue sky, it sent fingers of light to reveal the sleepers. Six young chickadees slept on a birch twig, clustered so tightly together that they looked like one long feathered animal, their collective length pulsating with uneven breathing. The moon shone on sleeping swamp sparrows and vireos, shrikes and thrashers, orioles and bluebirds, their heads thrust into their back feathers, their bodies fluffed to twice normal size to create an insulating barrier against the cool night air. All the birds slept with their body weight thrown back, which pulled leg tendons tight and clenched their claws firmly round branch or twig.

158

The eastern sky began to lighten. The whippoorwill cries died, and the cricket calls faded into the light and wetness of morning.

III

Many pond creatures possessed a peculiar quality of knowing which verged on intelligence. A parasite bumblebee flew in the entrance of a bumblebee burrow among the granite rocks one afternoon and scuttled underground to the honeypot. He was recognized instantly as a stranger. A dozen bumblebees pressed around him but did not attack or molest him. Then one moved forward and placed a drop of honey on the parasite's body. Another bee followed, and the intruder became agitated and scuttled back and forth among the bees surrounding him. His furry coat was now glutinous with syrupy nectar, but eventually he managed to push through the burrow bees and reach the surface. After several unsuccessful attempts, he freed his stuck wings and droned away sluggishly. The method of expulsion was puzzling, unless the bees had deliberately avoided a fight and possible casualties.

A day later, a deer mouse sniffed the burrow entrance. He smelled honey, and this drew him slowly down the tunnel. At the same time, a bumblebee entered behind him. The mouse became alarmed at the menacing buzz behind and bolted deeper into the nest, only to meet some bees emerging. A sudden flurry of violent movement filled the narrow tunnel. The rotund bees clustered on

159

the mouse's jerking body and jabbed in scores of stings. The animal finally burst to the surface, crawled into long grass, and died.

The formidable defense of the burrow did not intimidate a fly that, over vast ages, had grown to resemble bumblebees closely, though he had no sting. He even thrust out one leg stiffly ahead in simulation of a characteristic pose of an angry and defensive bumblebee.

One of these flies came over the pond one warm midmorning to the bumblebee burrow. He flew past an emerging bee, who ignored him, and landed near the entrance. He was immediately accosted by a guard bee, and the two creatures stopped and thrust out legs stiffly at each other. Another emerging bee flew at the posturing fly and tried to wrestle him into a stinging position. But the fly was elusive. The other bee buzzed indecisively. The noise of the fight brought other bees to the surface, but they merely blundered into each other and buzzed angrily. One latecomer suddenly seized another bee, and in a moment the ground was squirming with fighting bees. The uproar brought more bees pouring out of the burrow, and they joined in the fight.

From the scrimmage struggled injured and dying bees. The fly was lost in the fighting and was not seen as he skirted some struggling bodies and ran for the burrow entrance. He reached the underground honeypot in the burrow's chamber and immediately took a deep draft from it. Two worker bees examined him and then helped clean up his disarranged fur.

The fly would stay with the bees through that summer. He would not work and would have a unique status in the hive. He would witness strange fights between the queen bumblebee and other hive bees while she tried to lay eggs. The workers might eat the eggs, but she would persist. This determined behavior, coupled with the lethal hysteria of the fight at the burrow entrance,

seemed to suggest that the bees, lacking predators, had developed a system of controlling their numbers by themselves.

The fly would see the bees enlarging the hive by digging deeper tunnels and would watch them forming in a column, standing almost wing to wing, to pass grass down from the surface to insulate new underground chambers. The fly would eat and doze, and the bees' tolerance of him was a secret of the pond.

IV

The pond was incredibly fecund. The bursting green volvox, the budding youngsters of the hydras, the endless division of amoebae and paramecia, all produced uncountable millions of new lives each day. Floating masses of frog eggs hatched in thousands: droves of tadpoles specked the warm shallows. Hundreds of salamander eggs waited to release miniature creatures, perfect in detail, held inside transparent spheres. Freshly hatched turtles rose in scores from sand patches and ran for the pond. Those who were not killed before their shells hardened were lucky to survive long in the water.

The most spectacular multiplication of life in this hot season was the billions of algae staining the pond a deep rich green. They clustered on rocks and stems, and they fed thousands of pond creatures. The snails who grazed on them moved from one algae patch to another along midwater roadways of slime exuded from their bodies. The roads were reinforced by the passage of each snail. Occasionally, two snails would meet, and they might fight, butting shells together till one was knocked off the road.

This was a season of staggering numbers, of intense growth of life in every part of the pond. The milkweeds on the sandy peninsula harbored on a certain day a small colony of aphids clustered under a dozen leaves. Three days later, the aphids covered a dozen plants. Within ten days, all the milkweeds were infested with their globular green bodies, which were attached to the plant stems by sucking tubes. As they ingested plant

162

juice, they reproduced through parthogenesis, their off-spring swelling under their skins like buds, then bursting out as complete creatures ready to begin eating.

Their increasing numbers attracted a flood of predators. First came the aphid lions, hair-tufted, triangular-shaped and bowlegged, the larval form of delicate, fluttering lacewing flies. The lions moved steadily among the unheeding aphids, pulled their sucking tubes from the plants and drained their bodies, dropped the deflated, wrinkled sacs, and moved on.

As the lions fed, the lacewings hovered over them, seeking clear spaces between aphids and lions to deposit the eggs of more lion progeny. In doing so, they had to protect the eggs against the insatiable appetite of the lions. A lacewing found a small patch not thickly covered with aphids, and alighted daintily. From her abdomen she exuded a drop of fluid, which stuck to the plant. She pulled upward, and the fluid, hardening rapidly, stretched into a fine stem. She laid one egg on the top of the stem and floated away across the pond, where, in a flash of movement, she was snapped up by a swallow. An aphid lion came crouching through the aphid host and reared up the lacewing's stem at the egg. But the stem was too tall. The lion turned away and drained an aphid.

Other predators arrived to eat the aphids. Swarms of ladybird beetles and daddy longlegs moved into the milkweeds. Ichneumon flies, whose kind had shown their ingenuity in digging into wood and diving into water to find places to lay their eggs, hovered over the feeding

163

aphids and, with thread-fine ovipositors, injected tiny eggs into them. The eggs hatched, and as the aphids sucked sap, the grubs sucked aphids. The ichneumon grubs eventually cut small circular holes in the backs of the aphids and emerged.

Despite the carnage, the numbers of the aphids seemed undiminished. Their excrement, a fine, honey-like rain, attracted flies, ants, butterflies, and bees, who ate it from leaves and earth. The honeydew was so highly prized that some of the ant colonies secured domain over many of the aphids. They protected their territory by driving off other ants and coaxed honeydew from the aphids with caressing movements of their antennae. But finally, either the appetites of the predators were too strong for the aphids, or the force of aphid life waned. The multitude slowly dwindled. The ladybirds flew away; the aphid lions rolled into balls and spun silk cocoons the size of a small nut. After sixteen sunrises, small lids opened in the cocoons, and fully-grown lacewing flies fluttered out across the pond to begin again the cycle of life. Surprisingly, all the aphids were now emerging with wings, and they took flight and left the despoiled milkweeds to the pond.

V

A family of raccoons passed by the pond each night during this brilliant middle season. They usually arrived silently, their white extremities seeming to move disembodiedly through the trees, and their fondness for frogs and shellfish kept them hunting there for a quarter

164

of the moon's passage across the sky. The frogs feared them, and their singing stopped abruptly the instant the first raccoon was seen. In the silence that followed, a percussive chorus of frogs whacking onto the water rippled round the pond and a hundred eyes glistened from semisunken logs, lily pads, and from the water.

The manifold scents of frog, nymph, and shellfish brought the raccoons to the water's edge. One youngster lagged behind and nuzzled into the long grass. He

had found a wasp that had been struck down that afternoon in a skirmish with a robber fly. The wasp was paralyzed and, in trying to move his stricken limbs and wings, had only succeeded in pulsing his abdomen. The young raccoon, fumbling with his paw in the grass for the insect, failed to grasp him and so reached down with his mouth. As he touched the wasp, a sting was driven three times into his tender upper lip.

The raccoon stood fully erect, one paw held to his mouth. The prick of the stings was more surprising than painful. But then the pain started. It heated one side of his face as the poison moved to the eye, which watered and began closing. The raccoon tried to scrape away the burning with his paws.

He whimpered and ran to the others. They ignored him and went on hunting in the shallows while scuffling with one another. The young raccoon was nonplussed and ran along the shore, but the pain now suffused his face, nearly blinding him, and he blundered through some bushes and plunged over a small bank into the water. The other raccoons, sensing a game, came galloping to him and clustered on the bank, their pointed noses twitching. They watched the youngster ducking his burning head in the water, trying to wash off the pain, and they saw him come slowly out of the water, his lip swollen like a puffball and his left eye half closed. He sat down and began cleaning himself. The other raccoons moved away along the dim shoreline of the pond. Nearby in the thick grass, the wasp finally died.

VI

The pond exchanged a life for a life, or multitudinous lives for one. The exchange had been noticeable momentarily among the aphid-ridden milkweeds, and it had sent a spasm of new life through ladybirds, butterflies, lacewings, and ants. Another exchange was taking place along the curving edge of the pond's south shore. About forty toads had congregated there during the early evening, their calls screeching out monotonously. On this night, the toads collectively ate more than one thousand pond creatures—sow bugs, centipedes, beetles, leaf hoppers, May flies, caddis flies, midges, and caterpillars.

A pair of wrens caught three thousand small insects during the next day, to feed a nestful of youngsters. Two other pairs of wrens, also nesting at the pond, tripled this mortality rate so that more than sixty thousand insects were lost to the wrens every seven days. Yet the pond insects increased in number during that time.

The wrens' appetites were insignificant compared with the collective appetites of vireos, warblers, sparrows, flycatchers, phoebes, kingbirds, nuthatches, woodpeckers, flickers, creepers, and titmice. A pair of twitch owls, who were nesting in the nearby beech forest, every night sustained themselves and their brood on at least five sparrow-sized birds, a dozen mice or voles, perhaps a young rabbit, and scores of insects. Before the young owls left their parents, the pond and its area yielded them about one hundred and fifty birds, four hundred mice and voles, and a dozen young rabbits.

The owls did not hunt in the territory of the pond exclusively. They shared parts of it with two other pairs of owls and the red-tailed hawks, and transitory hawks and shrikes made numerous forays into the area. The pond's population nevertheless kept increasing.

VII

The pond was the center of a living universe that was dominated by the trees. The northern forest undulated unbrokenly for days of a bird's flight. Its mixture of trees had, through eons, developed an exclusive system of communal life that ensured a recreation of type. The forest mixed birches, beeches, and maples, forming an open-textured, high-canopied woodland. Each type of tree in it had distinct characteristics. The beeches, with their light gray bark and paper-thin, smooth leaves, were hardy in fighting for a place in the forest. They endured dense shade for scores of years but reached always for the leaf-mottled sun above them. Immediately after they approached treetop level, they spread a dense shelter of their own which discouraged others from following them.

Trees near them might be sugar maples, equally tenacious in their drive for the sun. They might contest with the scattered groves of hemlocks for a place in the forest. The hemlocks often excluded others with such dense leaf cover that even rain could scarcely penetrate it.

The forest soughed softly, and a huge white pine groaned in stately elevation. It was part of a magnificent stand of pines more than four centuries old which rose above the forest west of the pond. The white pines were opportunists, unable to stand shade and the competition of hemlocks, sugar maples, or beeches. They grew along the edge of the forest near pond and marsh, and if they penetrated the forest it was accidentally. An old beech had collapsed near the pond a hundred summers before and had brought down several other trees. With the canopy broken open, the sunlight had poured into the forest and new growth had sprung up to meet it.

Somehow, white-pine seeds had germinated there. Perhaps they had been carried on birds' feet or in animal fur or had been waiting, dormant, in the ground. Because they grew faster than all other seedling trees, they outpaced the others in the race to the canopy level. The pines now stood, close-packed and invulnerable, surrounded by birches, beeches, and maples. Much later, the big forest would overwhelm them, but for the moment they exemplified the infinite capacity of life to spread and survive.

The fall of a tree began a series of reactions. A large maple near the pond had dropped in high winds earlier in the year, and its new leaves lay dead around it. Its fall had brought a rash of plant growth out of the bare earth. Waterleaf plants had spread, covering most of the exposed ground, and their magnificent flowers spattered the new clearing. Enriching and encircling this surge of growth along the periphery of the clearing were yellow clintonias, purple trilliums, foamflowers, and partridgeberries; bellworts and disporum and round-leaved orchids, showing two prostrate glistening leaves with erect stems that carried spiked flowers.

The sunlit wreckage of the tree and the leaping new plants seemed suddenly primeval. A scarlet tanager stood vividly for a moment on a dead branch jutting up to the sun. A bobolink bubbled in the far distance, and the forest hummed softly with a legion of wings.

Another type of woodland grew south of the pond. It was composed of oak and hickory trees, which thrived on the warmth and sun penetrating the structure of the forests. Here was no suppression of competition but an invitation to light and air and a drying of already friable soil. Black oaks grew in the warmest, driest places, and their quivering, smooth leaves blurred the outlines of occasional chestnuts rising among them. The hickories could not survive in conditions that sustained the oaks and chestnuts—such as high, leached ground—but they were predominant elsewhere. Each tree was specialized to survive.

The shrubs were massed sleekly along the edge of the forest—mountain maple, dogwood, fly honeysuckle, and moosewood—preparing an environment for the germination of tree seeds. Aquatic plants encroached on the stream and the pond. They built up masses of roots in the shallows and caught silt there, inviting the land to move imperceptibly into the water. The western end of the pond, where plants grew particularly thickly, was most vulnerable to this insidious advance. Eventually, over hundreds of seasons, the plants would extend the swamp eastward, and new growth would appear that would consolidate the ground for the establishment of shrubs and trees. The great southern marsh had once been a lake, but the land, momentarily disguised as mud, tussock, and rushes, was annihilating it. The marsh might become a forest. The pond might not be even a hollow among the trees.

VIII

The lungs of the pond were the leaves. Billions of them hung, rustled, whispered, gleamed, and flickered. They sheltered the hunted. They fed insects, created shade, broke the force of the wind, and conserved earth moisture, but they did not reveal their most important function.

A fat green caterpillar was eating a maple leaf before transforming into a warbler-sized polyphemus moth. His chopping jaws worked regularly across the structure of the leaf and broke through the upper and lower surfaces, which were faced with an invisible cellulose skin. The upper side was smooth, the lower side rough, almost hairy. The upper side's transparent sheen protected the interior, and as the caterpillar chewed, he ruptured thousands of tiny interior cells. They gushed out cloudy, waterlike protoplasm, which dribbled from his jaws. Along the chewed line of the leaf, countless chloroplasts, the agents of photosynthesis, or food-making, could be seen inside each broken cell. Before disruption, these tiny green bodies had moved rapidly inside the cells, constantly uniting water and carbon dioxide to form sustaining sugars and starches. Each chloroplast contributed minutely to the tree's life, three billion of them in the caterpillar's leaf, itself one of about a million on the maple.

The caterpillar chewed through scores of liquid-carrying veins, or pipes, and ate thousands of breathing valves, or stomata, through which the leaf inhaled carbon dioxide and exhaled oxygen. The stomata were controlled by master cells, which opened them on sunlit days and closed them at night and on dull days. The unwanted substance of this activity, oxygen, flooded over the pond and into the lungs and lives of all creatures in an invisible shower of water expelled by the leaves. In one summer, the trees would release more water than was contained in the pond.

A mass of visible life ran through this silent leaf activity. Thousands of moth caterpillars ate the leaves in order to transform themselves into handsome sphinx moths, into huge cecropia and promethea moths, and into pale green luna moths with long, trailing swallow tails.

Leaf miners burrowed through the leaves, hidden within their thin substance but leaving exterior track marks of their interior passage. Leaf-cutting bees chopped out small pieces of leaf to line subterranean nests. Monarch caterpillars attached themselves to leaves and hung down in readiness for their transformation into pupae, and many were infested with grubs of tachinid flies, whose kind had become highly specialized in parasitizing other insects. One caterpillar, hanging over the pond, began to bulge and sway because of the infestation inside him. The grubs appeared through his hairy skin and lowered themselves smoothly on silken cords to the leaves below. Some cords snapped prematurely, and the grubs fell into the water. Dozens of tachinids waited among the leaves for their moment to dart forward and deposit eggs on the caterpillar hordes. Many caterpillars were so thickly studded with eggs that their green bodies were almost completely concealed. But for every thousand caterpillars that died, one or two survived and wrapped themselves in leaves and pupated.

The vivid life in the leaves appeared fully now. A sapsucker jerked into sight and stopped abruptly against a maple trunk, his bold colors—scarlet throat, red head, yellow abdomen, black and white wings—standing out sharply. He rapped a hole, watched it fill with sap, and drank. The leaves rustled, and caterpillars fell silently to the ground. Birds fleeted through high branches with beaks stuffed with insects. The leaves breathed out the life around them, and the sapsucker hurtled away.

cloudburst

Almost all the pond creatures sensed the disaster. Shortly after dawn—a red dawn of awesome proportions, stretching almost from north to south, bloodying the pond waters and eerily lighting the depths of the forest with a red glow—the pressure of the air began dropping steeply and registered in the senses of many creatures.

Birds feeding nestfuls of youngsters hurried in their hunt for food, and wasps and bees hastened from flower to flower. The noisy redwings were silent. The leaves hung limply. Some pond creatures had gone into hiding, like the frogs sunk deep in greenery and silence. The bullfrogs' eyes watched from the lily pads. The salamanders, dragonflies, and butterflies had disappeared.

As the pressure fell, the trees became a dark cavernous line around the edge of the pond. They were overtowered by the vastness of a dusky sky dominated by a black jagged cloud, the outer edges of which showed

174

brilliantly white from light behind it. This light seemed the product of some enduring explosion that was causing an incandescence as bright as the sun itself. In front of it, clouds moved rapidly back and forth as manifestants of sharply varying levels of pressure.

High above the pond, within masses of piled-up cloud, water vapor was rising and falling in great drafts of turbulence. At one point, the vapor was thrust up vastly, frozen into hail, dropped back into ascending warm air, melted, and thrust up again. Yet there was a perfect stillness at the pond.

The rain cloud consisted wholly of countless drops of water that individually were invisible but collectively appeared as fast-rushing wisps, columns, vortexes, swirls, and masses of cloud or vapor. The droplets were impelled to grow as water saturation in the cloud increased with the falling air pressure, and their collective weight grew so much that the updrafts could hardly support them.

Directly above the pond, a mass of drops of water began the long fall to earth. As they met warmer, denser air, they smashed into droplets and slowed down. The pond was shrouded under a thick mass of vapor that had not yet turned to rain. But the droplets would enlist some of this vapor in their downward fall. The rain gushed out of the lowest clouds, below which was the pond. Its surface was flatly gray, and its trees were stilled, waiting for the rain. No bird called; no life stirred.

As the rain fell, a conflict of pressures swept across

the pond. A violent wind roared through the trees, ripping loose elm leaves and slanting the rain sharply. The rush of wind and hiss of the water hitting the pond was eerie and desolate.

A group of chickadees, huddling insecurely in a big elm, were poorly shielded from the storm, and their protection was being torn away. The rain fled through the trees. It grew darker. The overcast pressed down, enveloping the pond. Some redwings tumbled hastily deeper into the reeds. Three juncos crouched together in a hickory, darting terrified glances into the noisy gloom. Three leopard frogs came bounding through the trees and leaped into the pond.

Underwater, the threat of chaos on the surface seemed to be communicated to frogs, nymphs, and salamanders, who scurried along the bottom and disappeared from sight, through water ominously darkened.

The pond now lay under a ghastly pall, and the wind bent the trees over the pond and sent leaf-clustered twigs and branches floating through the haze of rain. Finally, the great mass of water that had been held in the air suddenly overcame the forces holding it up, and the rain became an almost solid downward-rushing sheet of water. It struck the pond with a streaming hiss and chopped the surface of the water into foam, obliterated every landmark, pounded vegetation flat, filled the air with spume. There was chaos in the reeds, and drenched, dazed redwings floundered in the water. A duck flapped out of the reeds but was beaten back into the bulrushes.

In the roaring forest, a branch fell and a bluebird, dislodged from his perch, was driven toward the ground by the weight of the rain. He sought perilous refuge by clinging to the bark of an elm, where he was partially sheltered from the onslaught of the water.

The water poured through the forest in thick streams, flooding into hollows and holes and sending an astounded family of chipmunks leaping for their lives as their ground nest filled to the brim. The pond rose steadily, and the old muskrat ran agitatedly from the

central chamber of his shelter to see the water filling the entranceway, rising to where his mate was suckling newborn rats.

The inlet streams were roaring torrents, wrenching earth and vegetation from the banks of the sandy peninsula and spewing rubbish into the pond. The rain saturated all the ground, and brought worms pushing to the surface, where, half drowned, many were carried away in the torrent. An anthill was flooded, and the ants tumbled over one another to escape through the two

177

main entrances. Hundreds disappeared in the flood toward the pond and thence to the marsh.

Underwater, the flood entered the pond in rolling clouds of black, yellow, and gray scoured sediment. Many larvae and fish, unable to breathe impure water, were forced to the surface, and some died when they found no escape from it. The muskrat and his mate struggled through the water storm and surfaced on the shores of a disaster, where dead and living lay together in the incredible wreckage of their surroundings.

The flood had come from a sea halfway around the earth. It had been lifted into the air in the previous winter solstice, and it hit the pond merely because of a coincidence of pressures and temperatures there.

II

The rain was a prelude to something more destructive —hail. As the rain gushed down, the cumulo-nimbus clouds that generated it began changing their condition. Such clouds were hosts of thunderstorms and had a unique capacity for turning their masses of vapor into electromagnetic generators capable of storing great charges of electricity.

As the rain fell, this electricity collected, through some inexplicable process, and a series of great convection updrafts of warm moist air began racing through the clouds. These quickly condensed into more rain, but as the droplets hurtled upwards, they met freezing temperatures and were instantly turned into ice drops. Up they sped, acquiring coat after coat of ice, till they were each visible as part of a jostling quantity of balls of ice.

Simultaneously, lightning ripped from a cloud and revealed, in one stark second, a boiling prospect of multigrayed vapor in walls and canyons, pinnacles and valleys. The lightning burned through the air and created a huge vacuum, into which the vapor-packed air hurtled. As this air smashed into itself from all sides, it created an explosion that rocked the earth, and the concussion fled along the line of the lightning strike and ended with a crackle far beyond the marsh.

The noise of the thunder followed the upward flight of the hail, which was now losing its momentum. For a moment, it remained suspended in an air-borne sea, and then the updraft died and the hail began its fall. The hail balls were now as big as warbler eggs, and they soughed desolately as they dropped toward earth. The vanguard of the hail chilled the air and took its water and froze it and increased the size of each ball of ice. As the pieces of hail dropped through the lowest clouds, they were as big as robin's eggs and fell with a whistling roar.

The pond had prepared for disaster, but nothing could counter this onslaught. The first balls of hail churned the water, smashed through leaves, and bounced off the earth in a growling uproar of destruction. The hail brought showers of leaves, then small branches, then birds' nests and dead birds; but above all expectation, it brought itself in a mad, bouncing, ricocheting torrent.

The numbed birds could not cope with the storm. The nesting birds in the highest branches suffered most, and many died instantly. An oriole with a nest in an oak was driven off her eggs. She dropped with desperate speed and was nearly stunned by the glancing blows of the ice, but she blundered into a thicket and clung there, temporarily safe.

A chickadee was hit by a piece of ice nearly as big as herself and fell dead. A cawing crow fled through the trees amid small explosions of feathers as the balls of ice bombarded him. Scores of nests, filled with stones of ice, were smashed out of trees and fell, leaking pulped egg into the pond's conglomerate of ice and fallen debris.

As a massive overtone to this confusion, the lightning jumped from cloud to cloud, sizzled and burned into the earth and trees, and the thundering impact of its passage was almost continuous over the pond and beyond.

III

The convulsion of the pond was a small part of a greater storm. At its height, a weird screaming sounded in the north, rising and falling with terrifying intensity above the sound of the pond's crisis. Some unknown constellation of events in the upper air had started turbulence whirling in a gigantic circle, and unaccountably this whirling suddenly tightened, pulling itself into a funnel of air that screamed so loudly it drowned the noise of thunder.

The funnel rose very high and disappeared into a

black cloud that had suddenly become full of leaves, sticks, twigs, and vegetation. It moved slowly across rolling forested country, interspersed with ravines, small ponds, streams, and clear patches of ground, and wherever it touched, trees fell, branches smashed, and leaves were stripped away. In one glade, which the funnel was approaching, a wood duck waited apprehensively in a nest in a hollow tree. The scream of the funnel was muted in her deep refuge, but the sound was terrifying enough and she was ready to flee.

The funnel hit the glade as the duck appeared at the entrance of the nest, and she saw only a blur of leaves and debris before she was plucked away by a force that killed her instantly. Her feathers spewed off into the purple gloom, and her body, now high in the air, revolved in the funnel at tremendous speed, along with wreckage from the stricken earth. She rose with branches and stones, the heaviest of which moved to the outside wall of the funnel and were catapulted away, to fall in long curves to earth. The wall of the funnel looked solid, with its concentration of debris and vapor moving nearly at the speed of sound. But the center of its tunnel-like form, writhing up to a great height, was quite clear and was illuminated by flickering shafts of lightning that jabbed across it.

Directly in the path of the funnel lay a small pond, which was fed by the same stream that later ran into the larger pond to the southeast. The funnel trampled across the forest and into the pond, and in a few bizarre seconds the pond was gone. Everything was sucked

181

upward—mud, stones, water plantain, lilies, rushes, reeds, frogs, salamanders, a muskrat, and all the water of the pond, with its teeming summer life. Remaining was a muddied hole staring up to the sky. Meanwhile, the duck's body entered the funnel's black terminal cloud and joined an indescribable mélange of dead and dying.

IV

The rain stopped at the big pond, and soon after, a rain of another sort began. Frogs, salamanders, mice, and voles, all of them dead, fell from the overcast dome of the sky. They dropped unseen in marsh and forest. The wood duck, shapeless, almost pulped, splashed into the center of the pond. A long silence followed the fall of her body, and then a bird called hesitatingly. The clearing air was calm. The boiling cumulo-nimbus had disappeared, and the funnel had dissolved. The orange sun glowed from the western horizon.

The sodden earth was still absorbing the overabundance of rain water, and the creatures who had survived underground were trying to cope with the catastrophe. Many ant mounds had been flooded, and the workers clawed at glutinously blocked tunnels and hauled eggs and pupae to the surface to dry out. The earthworms who had not been washed away in the flood were moving slowly through flood-scoured tunnels.

In the shallows a large wasps' nest floated, humming softly with the activity of insects, who did not know that the entrance to the nest was underwater. As the heat mounted inside, some wasps tried to fan the fouling air and ran about agitatedly. The bullfrog surfaced near the lilies and looked across an unfamiliar pond of floating leaves, broken branches rising from the water, and dead bodies floating everywhere. He had kicked through a thick fog of silt settling slowly in the pond and had passed through many nymphs and larvae near the surface, all of whom were seeking the clearing water. Many had been suffocated by the silt, and as it settled, it took millions of invisible animals and plants to the bottom with it. It would be days before the rhythms of pond life re-established themselves.

The big bumblebee nest had escaped destruction in the storm, even though water had poured down the main tunnel and swept bees along with it. The nest was well founded and had drained quickly. Another smaller nest south of the pond had filled to its entrance and was only now draining slowly. The water-logged bees lay hunched in clots along the main tunnel, and the queen lay dead under the bodies of her workers. The great honeypot was still packed with sustenance, and some of the larvae, protected by waterproof cells, had survived the flood. But they would die. By the time they hatched, the honeypot would have been emptied by moths and other parasites of the burrow.

The storm had been most calamitous for the birds. It had struck at the seasonal peak of nesting activity, which, for the red-winged blackbirds in particular, was the most crucial time of the year. Few nests had survived intact, and one hundred redwings were drowned among the reeds. Birds who nested in hollow trees or under thick leaf cover were more fortunate, but in many instances the nests had survived but the parents had died. From a nest two young wood thrushes gaped blindly at rustling leaves above them, unaware that their parents floated in the leaf-clogged pond. The big family of chickadees, which the previous day had clustered like one long composite creature in the sun, was now gone. The chickens were scattered and dead, and their parents quested through still leaves, peeping inquiringly.

On a distant ridge of trees, a somber crow looked down into her nest and saw melting balls of ice mixed with egg embryos. She called several times, without getting an answer from her mate. Indecisively, she flew into another tree, and she looked down at the pond and its vale of destruction, but she did not comprehend.

Life stirred amid death. Throughout the forest beetles clawed from under leaves and passed into a strange world of caterpillars pelted down from trees, of aphids stripped from stalks, of cocoons and pupae scattered into waterlogged debris.

The mice came out of their burrows and looked, and like the crow, they could not comprehend the scope of the disaster. The forest trails were destroyed, and the familiar scents were gone. The varying hare, drenched and trembling, sniffed for a scent he could recognize. Nearby, his mate licked a dead young hare.

The late sun warmed the pond, and in the steaming moistness a notice of a new beginning rose from the soil. The rich earth smelled of purity, as though washed free of dust and memories. The birds stank with the heavy odor of damp and drying feathers, and the red-tailed hawk dried himself with a dizzying climb into the center of the sky. Dragonflies flared their wings in the sun, and flies and bees droned and buzzed free of the wetness. Later, the sun threw up a red shaft from below the horizon and caught the hawk descending.

The air was not free of moisture, despite the totality of the storm. When a bright moon rose, it shone through odd patches of mist, and it caught a patch of vapor at the eastern end of the pond. A moonlight rainbow trembled in the dusky air, showing clearly its four distinct bows. Then abruptly, the vapor moved on, the refraction ended, and the rainbow flickered out.

drought

The high-water marks of the storm dissolved into a final, brilliant efflorescence before summer began settling tranquilly into the fibers of pond life. The season was bright, blue, and stifling, the central disc of sun so hot it cleared the sky of clouds. The dry air sought to extract moisture from pond, leaves, and living bodies. It brought legions of birds to the pond, all trying to assuage the dryness that pervaded them, and the shelving margins of the water were spotted with bathers. Family flocks of goldfinches undulated to the water like showers of yellow and black petals being flung through the hot, still air. Birds waded into deep water and flapped their wings vigorously, sending up arching columns of spray. They ducked their heads underwater, waded, floated, and quarreled, and sometimes bathed so long they could scarcely lift their sodden bodies aloft again.

The water creatures watched the birds. Salamanders

paused in the shallows to peer at the stemlike legs coming down from the surface, then slipped away into deeper water. So intense was the birds' concentration on soaking themselves that they were often slow to react to danger. A sparrow hawk zipped over the pond, and a group of the bathing birds saw him belatedly and scattered wildly. Normally, the hawk would probably have ignored them, but he saw a purple finch laboring low across the pond in a desperate effort to sustain her sodden body in flight. The hawk pivoted back to the pond in one smooth movement and hovered over the laboring bird. Her wing tips were now hitting the smooth surface, leaving double circles of disturbed water, and then she was down in the water, still flapping feverishly, with the hawk hovering daintily just above her. The hawk hesitated to drop into the water, and as he waited, the finch made one final convulsive effort, which took her into the reeds, and there she burrowed like a mouse for safety. The hawk turned and sped off on his original journey.

The heat and dryness penetrated all living things, diminished hungers, and relaxed the urgency of life. A deer mouse sang a silvery trembling song in a moonlit thicket, and in the forest a spider purred an elusive song among the leaves. How subtly did the creatures of the pond communicate? Their manifold sounds now posed the question many times.

On perfectly still hot nights, when even the nighthawks were silent, a muffled whining sound rose out of the

earth near the pond and was heard by drowsy birds. The big bumblebee nest had prospered greatly during the summer. At its zenith, it contained more than fifteen hundred bees, and the narrow tunnel leading to the honeypot was jammed with workers sidling past each other from dawn to dusk. On three occasions the burrow had been extended by working parties. At night, this great quantity of bees sleeping together in the heat created problems of ventilation, and sometimes, after midnight, the air would get foul. As the heat rose, ventilator bees roused themselves and took up preappointed places along the entrance tunnel, with their backs facing the

188

surface. They buzzed without flying, and the corridor filled with rushing air and the drone of their wings.

The big white-faced hornets, whose large paper nest was fixed in a tree north of the pond, suffered in the heat. By late morning, the nest temperature would have risen unbearably, and first one hornet, then another, would begin fanning to push the fetid air down and out the nest's single entrance. Other hornets dropped to the pond to drink and returned to the nest with water, some to slake the thirsts of larvae in the nest, others to spread it on the absorbent exterior where it would evaporate and cool. But on the hottest days, their efforts were futile, and the interior became stifling. The hornets would evacuate then, rising in wind drifts to the tree tops. When the sun no longer hit the nest directly, they would return to it.

This was a time for lethargy, for dozing, and even for playing. A sparrow hawk swooped down into long grass on the sandy peninsula and hovered over a terrified field mouse, who dashed for cover in demented haste. But the sparrow hawk drove him one way and then another, without attempting to catch him. The game went on till the terrified mouse refused to be turned and ran blindly across some clear ground to a burrow. The hawk stooped daintily, never quite landing, and seized the mouse, whose squeaking cries diminished across the pond and dwindled to nothing above the southern trees. Perhaps the game was resumed elsewhere; perhaps the mouse died and was eaten; perhaps he escaped.

Even the wary turtles were imbued with the spirit of

the season, and in the hot sun, painted turtles crawled up a semisunken branch in the northern shallows. The younger turtles clambered over one another, climbed up on the shells of some adults basking there, and slipped and fell into the water and glided under the branch in short graceful chases.

While they played, other creatures were still preoccupied with breeding and survival. A red fog of silt rose in the water as some carp fanned clear spots to lay eggs on the bottom of the pond. The silt drove away all water life, and hundreds of hydras, who could neither swim nor float, were isolated in the fog, unable to hunt. Because they could normally only travel by a slow tentacle-over-head motion along plant stems or the bottom of the pond, their escape to silt-free water seemed problematical. They endured the silt briefly, then began moving. Each creature formed a bubble of air at the mouth of its suction foot, and each creature was carried to the surface by its bubble. The animals arrived singly, and then in scores, and as their bubbles popped through the water skin, they transferred their grip to the water skin itself. Soon the surface was a clotted mass of downward-hanging hydras, now free to hunt in the upper waters or to walk along the water skin's nether side to other parts of the pond.

The heat settled like a weight and evaporated the water faster than it entered the pond. Numerous creatures were caught in dried-out pockets around the edge of the pond as the water receded. In these dusty, mud-cracked basins, millions of invisible creatures survived as drought-resistant eggs, which might be blown to new territories or carried away on the feet of birds or insects. The receding pond left puddles of stagnating water, and in one a leech was caught. But he carried raccoon's blood in his stomach, and he would still be digesting it in the fall, if he could survive the drought. Elsewhere, planarian worms endured without food in small pools, but their starvation was sending them physically backwards through their lives; some of them already looked like newborn creatures. During this process of withdrawal, their vital parts atrophied, their sexual organs disappeared, their food tracts and stomachs shrunk. The worms would remain diminutive till the pond rose again, when they would quickly regain their lost organs and size.

In the stupefying heat, insects hunted diminishing supplies of nectar: some beetles ate flowers; spiders waited for night creatures in their webs; copperhead snakes moved silently through the darkness of bushes toward sleeping birds, which they could not see but which they traced by following emanations of radiant energy from their bodies. A soft scuffle in a patch of dark in the moonlight might be a puffed-up, gasping bird disappearing down a copperhead's throat. The nights were full of movement and death as skunks nimbly reached into nests to catch and eat young birds. Silent raccoons endlessly probed in the shallows for nymphs and frogs and one night alarmed a young fox who had been noisily drinking from the pond. He peered across the moonlit water with exaggerated wariness. This was

191

his first high summer, and he looked for enemies who existed mainly in his imagination.

II

For the first time, the odor of death became noticeable at the pond. Its sour smell mingled harshly with the rich scents of the forest. It threaded through the thick bushing hickories and pines, rose clear of the forest, and disappeared into the sky. It drifted up almost visibly from the clotted shallows under elms, where branches, reeds, and debris twisted together around the bodies of many storm victims.

The smell of death pervaded all levels of pond life, from the mud at the bottom of the pond to the tops of the white pines. Death sustained many creatures. The piping calls of a family of young nuthatches in a hollow pondside tree grew weaker and ceased. The male nuthatch had been killed in the marsh by a sparrow hawk, and the female had struggled to cope with her youngsters' appetites before being killed by an owl. Soon after the chickens died, a beelike fly was drawn into the hollow by the first faint smell of putrefaction, and her wings roared in the enclosed space as she settled to lay her eggs.

The heat in the hollow hastened the dissolution of the bodies and the hatching of the eggs into maggots. They ate into the chickens' bodies, submerging themselves in flesh already fast-rotting under the massive assault of omnipresent bacteria. The maggots faced suffocation in the almost liquid flesh, but they extended long tubes from their rear ends, pushed them to the surface to breathe, and continued their feeding, concealed.

Death was a process of reduction, and bacteria were the prime reducers. They were ubiquitous. They permeated the soil and floated in the air, clustered along stems of underwater plants, and drifted everywhere through the water. They were so small that no intelligence could measure them or describe their mysterious bodily functions or their exact appearance. Yet their contribution to the pond was supernormal, and none of its creatures could exist without them.

Some bacteria were eating proteins and producing ammonia and ammoniacal compounds, which were essential food to many plants. Some were eating the ammonia and transforming it to nitrites, and others were attacking the nitrites and reducing them to nitrates. The simple nitrates were the proper food of green algae, now scummily abundant in the pond, and much of the growth cycle in its water began with them. The nitrogen gathered by many plants would be imprisoned in their stems when they died if it were not for the bacteria. They attacked the dead tissue and released the nitrogen for all the pond.

Bacteria were in the eye sockets of a dead skunk, in the cranium of a dead crow on the northern ridge, behind the chitin of a dead dragonfly nymph buried in the mud underwater, in the collapsing rotten wood of a score of dead trees, and within the living trees themselves, eating branches and trunks hollow.

The water molds were close allies. A bluebottle fled

low across the pond one morning to escape a flycatcher and in one violent evasive maneuver hit the water and lay there buzzing and circling on his back till he died. His body was soon surrounded by a whitish outgrowing fringe of radiating fungus filaments. The molds ate through the body's interior structure and reduced it to invisible union with the water. The molds reproduced as they grew; each radiating filament ended in a tiny cylindrical sporangia that matured, was released, and became a free-swimming spore in search of new floating bodies. Some of these spores found dead desmids and diatoms, and others found a butterfly's wing. Even healthy fish injured in fights with turtles or other hunters were vulnerable to attack by the molds, often fatally.

Death in the pond was incidental and soon turned into new life. The insensible transmutation of one creature into another was a part of the pond's secret. Multitudinous entomostraca were scavenging amid the incessant rain of debris from the surface. These crustacean creatures, many so small they were invisible, ate almost anything—excrement, dead bodies, husks, seeds, leaves. They, like the bacteria, were purifiers and exchangers. Some, like the single-eyed cyclops, floated in vast numbers on the surface, as close as they could get to the heat of the sun. They sank when the sun disappeared behind a cloud, rose quickly to the surface when the sun reappeared. Without the entomostraca, the pond might be poisoned. Billions of dead and dying algae fell through the green water, having come to the end of their summer proliferation, and their bodies were eaten by the entomostraca. They were, in effect, being transformed into water, and their disappearance was as spectacular as had been their multiplication earlier in the year.

III

The enduring heat was burning the life out of the pond. It removed the color from the swamp's sweet flags and rimmed their leaves with brown, and it scorched the grasses everywhere. A rippling line of yellow asters fringed the swamp, and a veery's trilling, descending song sounded over them at dusk. The centers of the pond lilies shone brilliantly gold in the sun the next day, and a kingfisher flew rapidly along an effulgent purple-blue mass of pickerelweeds by the shores of the pond.

The trees were unaffected by the dryness because they were still drinking moisture from the cloudburst, but thousands of brilliant plants darkened into the full rich bloom of summer, and their growth slowed or stopped and their flowers shriveled into seed heads. But the memory of their colors lingered on in the heat.

Bloodroots had shown touches of delicate mauves and had been followed by the yellow of evening primroses. Gentians, bellflowers, and lobelias had matched varying greens with random touches of blue, yellow, and purple. These colors had been on the rim of the water, but now new and richer colors were transforming the pond.

On some evenings, the vivid sunsets stained the sky orange above the central ocher orb descending into a sea of flaming red. The color changed gradually into a deep blue-green as it ascended to the dome of the sky.

Bright colors of birds evanesced into the trees and were easily missed. A blood-red tanager paused on an alder branch, and an orange and black oriole perched in the shade of a tree. A goldfinch made a yellow mark against a black stone, and an exotic blue blur remained momentarily after the flight of an indigo bunting. A rose-breasted grosbeak left an impression of black, red and white as it stood alone for a moment on the white sand of the peninsula.

The colors became subtler and warmer as they emerged, in all their variety, from the heart of the pond. The bluish-green needles of the white pine rippled around the reddish-brown seed cones, and graceful nodding pyramidlike hemlocks mounted to delicate peaks of rich dark foliage that glittered in the noonday sun. Visually, the pond was now endlessly various, showing white birch bark partially hidden behind foliage that rustled and chafed noisily in hot puffs of wind. The trembling aspens held out stiff leaf stems that caused the leaves to turn sharply in the wind and mirror the sun.

Tall milkweeds, long since recovered from the aphid invasion, thrust up thick clusters of red and white flowers from the soggy marsh and around the low shores of the sandy peninsula. This upwelling of sap in the drought attracted nectar hunters, and at midday ten thousand of them—wasps, bees, butterflies, and many kinds of fly —climbed the plants' stocky stems. A bumblebee flew over the marsh and dropped into the middle of a mass of milkweed flowers, digging his long proboscis into

196

the flowers and bumping and scuttling from bloom to bloom with the speed of gratification.

The milkweeds concealed a secret. As the bee blundered over the pinkish white flowers, his feet often slipped into some of the five narrow slits in each flower. In his hurrying quest for nectar, he impatiently pulled his legs free from the momentary clasp of the slits and rushed on to the next blossom. But as he pulled, a concealed pair of prongs, tipped by an orange square of waxy pollen, clamped at the tips of his legs. Sometimes, the pollen tips kept their grip and emerged with the legs, and the bee shortly had pollen attached to four of his legs; but he ignored it and kept bumping from flower to flower. Even with these attachments, his legs still penetrated the mysterious slits, and so their purpose was revealed. Often the pollen stuck to the stigma at the top end of the slit and fertilized the flower. Around the bumblebee were thousands of bees and wasps repeating this process.

The complexity of this process of fertilization momentarily obscured the fact that in this hot and tranquil season of sun and nectar, the milkweeds were a puzzling contradiction. Hundreds of insects had not had the strength to pull their legs free from the slits. Ants hung, dying and dead, from the flowers, many of them biting furiously at the nectar-filled petals now thrust into their faces. A monarch butterfly flapped and twisted and turned in desperate efforts to wrench free. A crane fly that had simultaneously straddled three milkweed flowers found herself in a strange predicament. Her long legs were caught in two flowers, and the wind was jerking the flowers back and forth so that at each jerk she was stretched out tautly. Eventually, a vigorous burst of wind ripped both imprisoned legs from her body, and she flew off across the pond with her remaining four legs dangling. Scores of flies buzzed and roared with blurred wings in their efforts to escape from the traps. This mass execution in the swamp did not discourage insects from hunting the nectar nor did the hunting damage any of the flowers. The final paradox was that the trapped insects did not fertilize the flowers. Through a quirk of natural history, the milkweeds squandered this life for no discernible reason.

IV

The rain had to fall eventually, and through hot days new masses of cloud formed above the pond, summoned from the hot vacuum of earth. The sun poured down on clearings and rocks, and the heat bounded back into the air and helped create cloud. Directly between pond and marsh, very high, a cumulus cloud had remained all one afternoon, and a steady breeze in the upper air did not move it across the marsh. It constantly regenerated itself from the rising warm air and just as quickly degenerated and died on the side leeward to the wind.

The cloud, which disappeared shortly before evening, was the only visible sign of the complex interplay of forces in the sky above the pond. This was a region of surging updrafts, of slow whirling rolls of warm air colliding with falling cooler air, of shafts of wind lancing directly through clouds. This interplay was a tiny part of enormous movements in the unseen universe, movements that stretched coherently halfway around the earth, some having their genesis over deserts, some surging into life in the southern antipodes, then in darkness.

The clouds coalesced along the horizon, thick white and dense. They grew upwards and leaned over the pond, and the sky became bluer as they climbed, or perhaps it only seemed so because of the contrast with their whiteness. Their gigantic, towering size and endless, castellated substance changed all the dimensions of the pond. Their whiteness was unblemished till a tiny speck of life moved across one of them. It was the red-tailed hawk, riding on a towering updraft between two clouds. To him, the air was not an invisible void but a substantial, palpable firmament that moved about in predictable and useful patterns. He knew exactly where each surging column of air rose from the ground, and he moved from updraft to updraft, coasting steadily lower between each lift, then rising high again. Sometimes the force of the updrafts was so strong that his wings would bow and the tips would quiver. He turned back and forth now between two silent white buildings of the air, then caught an updraft that sent him up at great speed. He saw the pond briefly between the two clouds, then he was alone and passing over a cloud and sliding down toward the marsh.

Spread out before him was the smashed reflection of

the sun shining back from the great lake on the horizon, and the vast slashes of forest encircling small, green lakes. He saw the weaving lines of streams, brooks, and rivers moving toward the great lake and seeping into the marshlands. He was now excited by the proximity of the clouds and by the strength of the supporting air, and he screamed for pure enjoyment of that moment of flying, even though, being so high, he could not hunt or even clearly see any prey on the ground. He cut through the air with sculptural precision, shaping a course around an outflung turret of vapor, his wide amber eyes looking down and his wings spread tirelessly against the moving skin of the earth.

V

Evaporation was invisible in the heat. Billions of droplets of water rose continuously from the surface of the pond, rising faster and faster till eventually, the product of a complex, mystifying aerial process, the droplets became clouds. The power of the sun to draw the water into the air was, however, also the power that sent the moisture whirling around the world in winds so high that the pond never felt them. In this way, the skies were kept clear, blue, and hot, and the stifling drought moved to its conclusion.

The blue sky seemed clear. On cloudless days, when its brilliance shaped the sky into a shield that enclosed the pond against the universe, a grand illusion was at work. The blue was not caused by clarity but by turbidity,

or countless dust particles. Each space of air the size
of a robin's egg contained more than a million of these
particles, and they were filters that reduced the sun's
heat and cut out the reds, violets, and greens of light
from space, allowing only the dominant color of blue to
reach the pond. The winds could blow and the rains fall,
but the eternal pall of dust remained over the pond. As
fast as its particles fell to earth in rain and dampness,
more were swept up from dry parts of the world.

A similar sort of pall existed in the pond. But whereas
the dust particles in the air remained suspended, the
particles in the pond fell continuously. They included the skeletons of invisible animalcules and beetle
droppings, globules of sap and dragonfly wings, bird droppings and pollen. The surface of the water was constantly catching air-borne debris and temporarily holding it
on its taut surface skin, and the pall of the pond filtered
the light of the sun till at the bottom it was deeply
blue-green.

Above the pond, the red-tailed hawk balanced on an
updraft. Below him flew a marsh hawk. Below the marsh
hawk flew, clumsily, a group of praying mantises who
had been caught up from the earth in an updraft and
were now moving rapidly east. Below them were groups
of nameless flies bound for destinations beyond the
pond. Below them were millions of flying aphids who
had been migrating from some milkweeds and had been
caught in the thermal movement over the pond. Below
the aphids was a speeding sparrow hawk, and below the

sparrow hawk, now at tree-top level, were darting birds of many kinds. Below the birds were flower-hunting insects and flies, and below these was the earth-bound varying hare, watching the three hawks, and also one of his half-grown progeny on the other side of the pond. Below the hare were the ants and beetles. Below them were frogs, salamanders, fish, and worms, and below these were aquatic nymphs, and below them were the buried shellfish at the bottom of the pond.

The layers of life might unexpectedly rise from the depths or fall from the heights. After the moon rose that night, it reflected from the pond. This was a signal for a host of flying creatures to close their wings and fall. They were sharp-edged and almost flat, and they dropped as fast as stones, hitting the water and disappearing in ripples. The thickening stream of their fall stopped the instant the moon went behind a cloud.

Beneath the water, they appeared as diving beetles, refugees from other overcrowded ponds. They had sur-faced, and at the moment of emergence had lifted ar-mored covers and revealed transparent wings. These had taken their sheeny black and purple bodies high into the night air. The beetle migration was general, rising from many ponds, and the insects flew across forest and marsh while waiting for a glint of moonlight on water to send them plunging down into a new submarine life.

The phenomenon of creatures who had wings solely for one spasm of their lives was visible briefly elsewhere in the imperceptibly waning heat. In many ant mounds on northern slopes and in the southern forest, ants with wings were now waiting to emerge into the flying life of the pond. They resembled their fellows exactly, except for their wings, and their presence seemed to agitate all the ant-mound dwellers, particularly on warm sunny days. At one moment, all the winged ants began moving to the surface, hustling up dark tunnels and jostling unwinged ants in their pre-eminent purpose to reach the hot sunlight. They tumbled out of the mounds and rose quickly into hot puffing upper air. Later that day, and on ensuing days, male ants fell steadily from the sky, dying and dead. Their lives were ended the moment they mated with the flying females, who were themselves coming to earth to face a new beginning of life. With the living sperm of the males within their bodies, many would now remain pregnant for ten to fifteen years and then produce colonies of their own creation. These new ant communities would all have sprung from a few hours of flight in the blazing summer sky over the pond.

the voyagers

After light rains broke the drought, the pond changed abruptly into the next season. The change was discernible one morning soon after dawn before either birds or insects had begun singing. A mist-laden pall blurred the extremities of the marsh and spread across the pond and became a dimensional, silent thing, spreading into a vacuum left by a departed torrent of noise. The moments passed interminably, and still no morning songs of birds sounded. There were no frog calls, no burring crickets or cicadas, no restless surge of life visible anywhere. The long, thin silence stretched ahead into the emerging day.

This transformation was audible and atmospheric. The air of the pond had changed, and a thistle seed floated infinitely slowly over the water in a firmament of deep tranquillity. The pond was suspended in time and space waiting for its change of life to quicken into the next season. The call of a bobwhite quail from the north had in it a touch of melancholy which foretold that this was the time of departure.

The fractionally shorter days were cooler, and this atmospheric change invisibly began stopping the photosynthesis by which plants and trees formed sugars and starches from sunlight. The chlorophyll green was slowly leaching from the leaves and would soon expose pale yellow pigments like carotene and xanthophyll, previously hidden under the summer color.

The new season exposed itself starkly one night when the temperature dropped far below normal. Many flying insects remained in their nests or stayed immobile on leaves and grasses. The wasps crawled from their nests on this chill day, but they could not fly. Some flies took to the air and droned sluggishly and clumsily through the trees. In the evening, thousands of moths tried to launch themselves into the night, but though they beat their wings stiffly, none took off. The low temperature had numbed the high-activity muscles of the insect fliers.

Though it warmed the next day, the pond had already begun a vast retreat which would be at first hidden in soil, wood, and water but which would soon become perceptible. Myriad ants had become sluggish and retired deeply into their mounds to prepare for sleep. Still-active ants climbed over them in corridors and galleries on their way to forage on the surface, and others dragged pupae deeper into the nests. Many pupae were transformed into adult ants and immediately went to sleep;

their arrival at the surface would be delayed two hundred days.

The sloth of the ants was one reaction to the season, typical of many insects; but the animals responded with new vigor and determination, which propelled them out of summer languor in an almost feverish hunt for food. Frogs leaped wildly through the grasses, and families of

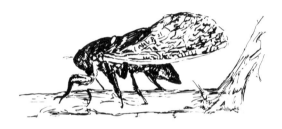

grouse walked urgently along the pond's edge. At twilight, the skunks and raccoons hunted without playing, and a wavering loon call from the marsh accompanied the first sounds of birds traveling high and fast in the night sky. The moon caught a glint of gold on a beech leaf. This was going to be a radical season, reacting with disruption and death to the end of a season of creation.

II

A chill penetrated the pond and its multifarious fleas responded: the females produced the first male water-flea eggs of the year, which hatched and immediately mated with the females. This began the formation of larger and tougher eggs that could endure the long wait on the bottom of the pond for spring.

Overwintering eggs, supreme examples of how evolution had shaped life to surmount any difficulty, enabled many dissimilar creatures to survive the winter. The fragile sponges had lived through summer as complex structural galleries and corridors of tissue that ingested regular cycles of water, from which they filtered food. Their delicate structure was clearly incapable of surviving freezing and crushing in ice, and since they could not move from their anchorages on rocks and stones, they needed a radical change to endure. Deep within each sponge, tiny spherical gemmules, or overwintering eggs, jostled together in the incoming streams of water. The gemmules contained food to sustain them through the winter, and as they bobbed about, they were forming tough, ice-resistant coatings. Each was minutely pierced at one point. Through this hole would escape fertilized eggs, which would start forming new colonies of sponges in the spring.

The overwintering eggs were often mobile. Down the inlet streams came a flood of statoblasts, which were invisible buds of cells from bryozoans now beginning to die in streams and ponds to the west. They trickled into the pond so thickly that they floated in visible brown masses, and many caught onto aquatic plants by lines of hooks around their chitinous shells or sank to the bottom. The spring growth of bryozoans would be spectacular.

Black crickets forced their flexible abdomens deep into patches of friable earth on the peninsula of the pond and laid their eggs. Katydids fastened hundreds of flat oval eggs along the edges of leaves and twigs. Walking-stick insects pumped a rain of eggs from trees to the ground, where they lay encased in brown, almost transparent skins, awaiting a protective cloak of leaves and snow.

In trees and among plants, scores of millions of plant lice were already concealing eggs in bark and plant fiber, and leaf and tree hoppers had hidden eggs in grass, which would melt into the ground with the rotting vegetation of winter. Aphids, laying eggs in tree bark, were providing a plentiful winter harvest for nuthatches and chickadees, any one of whom might eat five hundred of the eggs a day.

A mantis climbed the stem of a bush on the south bank of the pond and hung head downward. She squeezed a mass of white material from her distended abdomen, and tiny beaters at the abdominal orifice whipped the material into a thick froth. She moved her tail from side to side to build up the froth into a precise shape while simultaneously laying eggs in it. As soon as the last egg was ejected, she flew off and was gone forever. The egg case hardened and darkened into a tough horny mass, and the eggs were safe, at least till spring.

III

It was now the time of the travelers. A thin stream of monarch butterflies came out of a clear sky late one afternoon. They appeared over the trees singly and in straggling groups. At times the air flickered black and orange with their massed beating of wings. All were headed for the vale of the pond for rest. They settled in the trees, clustering in clots on bushes and in plants. One hundred of them settled on one small stem, bent it over, and stained the pond water with their wings as they died. The butterflies slept at the pond and were gone in the first flush of dawn, heading into a pink southern sky for marsh and great lake and the curve of the horizon.

These were days of damp expectancy and sudden winds and drizzle. Even dragonflies were flying south, perhaps migrating like the butterflies. With them flew some wasps and red-admiral butterflies, on uncertain migrations of unknown duration. A thousand dragonflies moved high and fast overhead one midday and flew over hundreds of dragonflies at the pond. Why some dragonflies migrated while others did not was not explicable.

The insect migrants revealed how common laws regulated all life, because other creatures, unrelated to them, were identically absorbed with escaping from the north. Traveling birds appeared in the air and trees. More than forty hawks, mostly broad-tailed and sharp-shinned, passed over the pond in one day and went on across the marsh. The red-tailed hawk was fixed in the sky one day and was gone the next.

The pond was a jumping-off point for many birds unsure of the distance across the lake, and they lingered around its shores so that the migration grew steadily noisier. On one cool day, more than ten thousand blue jays flooded past the pond: their massed flight was overwhelming, as though a single creature of unbelievable size had exploded into sight. The noise and movement of the jays obscured hundreds of quieter migrants like the ruby-throated hummingbirds and thrushes, who lurked in thickets while a flock of the jays climbed almost vertically till they were visible only as dots and then dived with closed wings and roared, screaming, into the trees.

Above their uproar, high and almost silent in the sky except for odd calls, thirty thousand crows passed in one day. The next day, another twenty thousand passed, and forty thousand the next day, and twenty thousand the next day, and then fifteen thousand and thirty thousand and fifty thousand, and black columns of birds stretched almost unbroken from horizon to horizon.

A thousand robins arrived and spread along the grass margin of the pond and through the marsh. One bird left the flock and flew over the peninsula, past the swamp, and settled in a tree near an old nest resting on a branch there. She flew to its crumbling rim, flipped out some leaves with her beak, then sank into its cup and turned around as though preparing to nest again. Abruptly, she left the nest and returned to the flock. Was this some incident in the robin's memory being re-enacted before she left the pond forever? In the late afternoon, the robins rose together and flew south.

The pond saw the gathering of birds from the huge width and depth of the north. It was on the path of a migration route along which traveled, in one day, fifteen thousand terns and twenty-five thousand chickadees, bald eagles, and pigeon hawks. This was a convulsion of natural history. Twenty evening grosbeaks broke their journey to drink dewdrops that hung from pine needles, and a kingfisher paused to rupture the center of the pond with a dive while the travelers passed steadily overhead.

The startling flight of the jays was soon surpassed by the gathering flocks of red-winged blackbirds. The resident redwings at the marsh were being joined by thousands of newcomers pouring down from the north, and soon there was such a congregation of these birds that they stained the gray skies black as they circled and climbed in sinuous patterns, rising like the jays to great

211

heights, then spreading out like black water over pond and marsh. One hundred thousand of them roared overhead in a singly motivated mass, and on one occasion two flocks gathered at opposite ends of the marsh and headed inland toward the pond. They climbed steadily, and both turned toward each other, the flock leaders aiming diagonally down at the pond. At tree-top level, the birds flattened out in a roar of wings and hurtled together in chaos. For long seconds, the air above the pond chattered and gasped and squeaked as the flocks mixed; then with unexpected suddenness, they were gone. One purple-black feather drifted slowly down to the water. A woodpecker, hushed during the spectacle, rattled at an elm branch. A bobolink tinkled distantly.

While some birds played or rested at the pond, others swept past it. The lines of geese in the sky did not waver as they approached the wide water ahead and passed on, their leaders honking powerfully. The pond frequently resounded to the sharp stutter of short powerful wings pumping quail overhead.

Twenty thousand tree swallows came by the pond in a rush of sound. Their vanguard rose and spiraled almost out of sight, as though to vault across the water. They dwindled and disappeared over the great lake. The variety of migrants seemed endless. Bluebirds, golden plovers, sparrows, and stilt sandpipers passed in groups and alone, by day and night.

The pond was a witness and a provider. Migrating robins paused to eat the fruit of mountain ash, honeysuckle, and haws. Crops of nearly all the forest seeds and fruits sustained the migrants briefly. Rowanberries, mountain-ash berries, winterberries, nannyberries, hemlock cones, and hazelnuts were plentiful. Goldfinches picked seeds out of spruce cones, and all around the pond birds were stuffing themselves with food for journeys that might stretch halfway around the earth.

In the midst of this southern movement, a group of flying creatures arrived from the south. They appeared over the trees south of the pond as countless tiny specks. Two indigo buntings, taking off from the peninsula to head south, turned quickly out of their way. The floating specks came closer, and the sun sparkled on thousands of strands of floating silk, each one ending in a suspended spiderling. A downdraft brought them over the pond, and silk caught in branches and swung some of the spiderlings into trees. As they touched wood, they cut the strands and scuttled away.

Most of the spiderlings were less lucky. They dropped to the water, and once their silk strands were below the influence of the tree-high wind, they remained in a lea and slowly sank. Some saw the danger of the water and climbed up their silk, but this shortened their flylines and sent them falling faster.

The spiders were soon dropping to the water skin, infinitesimally wrinkling it; and elvers, frogs, nymphs, and salamanders, carp, bass, and sunfish ate them by the thousand. But though the spider flight had been obliterated from the pond, the few survivors in the trees were sufficient for survival. Before the last of the spiders had died in the water, the survivors were seeking places for their hibernation in splits in bark and in hollow branches, readying for their adulthood of next summer.

IV

The aerial travelers passed on, throwing their shadows over other types of voyager proceeding with identical resolution to winter refuges. The aerial creatures traveled across the world to find safety, but these other travelers humbly merged with the earth and found it there.

This underground migration had already begun in the pond, though the air had scarcely chilled its waters. But the migrant creatures measured the shortening days. Scores of frogs were about to burrow into the thick underwater mud. Some dragonfly and damsel fly nymphs would seek heaps of vegetation or hide themselves under waterlogged branches, and water beetles would become sluggish and drop to the mud to begin a stuporous half-sleep there.

The pond offered shelter to many. From the northern wooded slopes came a steady stream of leopard and pickerel frogs who had spent the summer foraging among the leaves and long grasses for grasshoppers and other insects. They jumped into the pond and remained floating as if contemplating the sudden change from hunting in the open forest to sleeping deep in mud.

Their migration nearly coincided with another, moving in two directions at once. The adult red newts, who had lived on land for two years, began moving back to the pond. Some came from great distances, and as they wriggled overland, they passed many young newts recently transformed from gilled larvae into lung-breathing pond creatures. They were leaving the water and moving into the forest to begin their two years on land, which they would spend sleeping in leaf cover.

The drive of migration was to water and soil. Countless leaf hoppers, ground beetles, and aphid lions headed down stalks, stems, leaves, and trunks for the protective soil. Once into the soil, they squeezed through cracks, dug into soft patches, squirmed into worm holes, slid under leaves and rubbish. The forest leaf cover absorbed millions of them, and in thick layers of humus and leaf, rotten wood, new leaves, and rubbish they found insulation and a chance to create their own miniature climates, which would protect them from freezing. Beetle larvae, spiders, plant bugs, worms, and springtails moved deep into the leaf cover while another flood of travelers—snails and spiders, long-legged harvestmen, beetles, and fireflies—dug into rotting branches and trees. Torpid clusters of flies, many of them groups of females that were already fertilized in readiness for the spring, clung together in deadwood and tree hollows.

For some, fall travel was involuntary. A hundred snails, fastened to the stems of pondweed, fell with the dropping plants to the bottom, and under them other snails were heading for deep water to escape the surface ice. Many snails had withdrawn completely into their shells and were sealing the entrances with a tough, thin cover that would retain body moisture and repulse the cold. The great pond snails glided over mud to the deepest part of the pond, and two of them oozed over the edge of a stone and fell, slowly turning over and kicking up puffs of mud where they struck bottom. Within moments, they were half out of their shells and moving steadily down. Some small snails migrated short distances in the shallows, pushed into leaf piles or edged themselves under stones, and five hundred spotted newts swam up the pond's gravel-lined outlet stream to the marsh and settled to the deepest mud.

The onset of autumn warned water boatmen and back swimmers that they must migrate or suffocate or freeze in shallow water. They entered the big pond steadily from small ponds and shallow marshlands, and most of them were flying against the pale blue sky for the first time in their lives. They pattered into the pond and pushed out of sight among the wilting waterweeds.

All creatures responded to signals, perhaps of temperature or of light, or perhaps of indefinitude. The pond toads, who had lived along the banks of streams and around the sandy peninsula and in swampy ground to the west, began moving purposefully eastward late one afternoon. Migrating birds, resting in trees, saw hundreds of them leaping along the northern and southern banks of the pond, headed for low moist bush-studded ground lying between pond and marsh. While the last of them were still passing the pond at dusk, the first of the marchers were digging themselves into the ground.

They used their powerful back legs to kick the soil from under them, and they sank down rapidly. Within two days, all the toads of the pond had disappeared into the earth.

v

The first frost killed millions. It was a preordained death, not a disaster. Many monarch caterpillars which had not been able to pupate in time to join the southern migration died and dropped out of the trees during the first frosty night. The frost also killed much vegetation around the pond. It wilted leaves and stems and sent water plantains toppling back into the water. The cattails remained upright and green, but they were dying. The lily leaves disappeared one by one, folding up and falling back into the water. The duckweeds sank and disintegrated. In the forest clearings, the thick grasses of summer, long since dried out by the drought, were mostly dead, and their late vagrant green leaves wilted back into the debris of summer.

In the midst of death and retreat, there were denials of the new season. The dark purple berries of the pokeweeds sustained many forest birds for the drastic new season to come. The dogwoods, with their profuse red fruit, would feed thousands of birds, and many migrants ate blackberries, which grew in stunted clumps on the peninsula. Foxes, chipmunks, and squirrels ate red chokecherries, red bunchberries, and viburnums; and with others, they also ate acorns, beechnuts, hickory nuts, winterberries, bittersweet, and smooth sumac fruit.

The pond's wasps and bumblebees had become steadily more sluggish during the onset of the chill. Earlier, the queen bumblebee had been pursued by a worker and had mated with him in the grass. Now she flew out of the hive to look for a sleeping place for the winter, and she left behind her the fifteen hundred workers she had bred there. Their coats were now tattered, contrasting with their sleek splendor of summer, and they were lethargic, some dozing in the burrow, others clustered around the entrance in the weak sun, moving slowly and aimlessly. As the days passed, they left the burrow and found places to hide in the forest. Some crawled under leaves; others disappeared into holes in trees or into long dead grass. This seemed less like dying than like a life force gradually expending itself.

The big paper nest of the white-faced hornets which hung in a beech tree now contained forty queens. The males had disappeared into the forest and were dying there. In the cells of the nest were nearly two hundred pupae about to transform into male adults. They began emerging, and within a few brief days they had mated with the adult queens. The queens then flew into the forest, and the males remained sleepily clustered in the nest. Many would leave and scatter into the foodless and inhospitable world around them. Some would remain in the nest, as though reluctant to concede to such an early end to their lives, and would still be alive when the freezing weather began.

The queen hornets had found a score of hibernating places. They had bitten into the pulpy remains of rotten tree trunks till they had each excavated a small chamber. Over ensuing days, they gradually assumed their stances of hibernation, legs stretched straight back and wings nearly enfolding their bodies. The bumblebee queens had returned to the rocky outcrop and were huddled into dry corners, heads down, legs pulled under their stocky bodies. The carpenter bees, which had extended their corridors through a dry elm branch during the summer, had returned to the corridors where they were raised. They now lay asleep, end to end, in the same position as at their birth, but instead of facing up the tunnels, they now faced down them toward the spring that would awaken them half a year hence.

The travelers passed on—flying, crawling, leaping, swimming, burrowing—and at night a luminous band of stars stretched from northeast to southwest across the sky like a running beacon above the lines of migration. It divided into two parallel lines and remained divided till it disappeared into the southwest.

Bright Polaris, the north star, shone on the withdrawing life. Chained Andromeda in the east and Hercules in the west flanked the fleeing creatures. The Northern Cross glistened remotely.

CHAPTER 13

the golden forest

The trees turned their greens to red and yellow and became blobs of vivid multicolor irregularly surrounding the pond. The reeds were now a drily rustling slash of gold rippling across the pond's south bank, and as the trees were washed with vivid new colors, the pond lost the last of its subtle tints. The blues and greens of its water life were gone, and the almost clear water reflected the spectacle of the shorelines.

The light of sunset suddenly became more intense, and the pond colors grew fulgent, almost unreal, as the sun went down and the last of its thick light splashed along the shore. The pond seemed to smoke and sway before the sun disappeared, and the colors with it. A late redwing chucked, and the gloom came up out of the trees like death, and the trees blackly bisected the luminous purple sky and the passive water.

The changing colors persisted through graying days

and isolated stabs of real chill. The birches turned a pale yellow, which was spaced by the dark marks of green conifers and enlivened by the richer colors of the beeches, which, like the birches, were losing their chlorophyll but at the same time were stimulating their yellow pigments with dashes of brown. This gave the yellow touches of gold and surpassing richness.

The fall of the leaf was one of the many events at the pond which seemed simple and inevitable but which was the product of eons of evolution. It was not enough for the leaves to die and break from the trees; this would create no clear division between the living and the dead. It would be equally impossible for live leaves to break spontaneously from the living fiber of the tree. Like all living pond creatures, the trees were sensitive to injury, and during the great storm every leaf torn away left a bleeding, unprotected stem vulnerable to the entry of bacteria and funguses.

The trees were now forming a barrier at the base of each leaf stalk which would stop sugars, formed in the leaf, from circulating into the tree but which would not block sap from reaching the leaves. This flow of sap stopped the leaves from wilting and triggered the leaf sugars into a series of physical changes that spattered the pond with dazzling colors.

When the stalk barriers were completely sealed, the leaves began to fall, starting their second seasonal phase, the enrichment of the ground with potassium, phosphorus, magnesium, and nitrogen. They would be eaten by bacteria and funguses and so provide the trees with nourishment for other seasons of growth. The dead leaves, like the animals who had died in forest and pond before them, would eventually be transmuted into living things again.

These were days of restless winds that rushed suddenly through the trees and whipped the water into breaking wavelets. But the winds went nowhere and left in their wake sudden pockets of calm. One moment the leaves would be flying over the pond, curving down in skeltering masses, whipping up to the sky, then swooping to the water to stop abruptly and spread out like yellow and red stars. The dirty white skies showed patches of blue, and through the agitated air came the first real notice of winter, barely distinguishable snowflakes that floated down and disappeared before they could land on earth.

Suddenly the flying leaves were joined by a flapping mass of grackles, purplish-brown birds with sheeny feathers, who brought a comradely spirit of life to the pond. One dived to a grassy patch north of the pond, and soon the ground was speckled with the quickly walking birds, all of them darting, pecking, fighting, and chackering. At any moment, some would be in the air, some in the trees, some flying back and forth with important purpose. The snapping of a twig sent them rushing into tree tops, but before the last of them had settled, some were returning boldly to the ground. There was chaos in the grass at their coming, as aphids, grubs, and worms were ripped from burrows and hibernacula; and a thousand sleeping insects died in moments amid tremendous din as the clattering birds filled their stomachs. As the dusk settled, the birds rose silently and disappeared almost as though some common signal animated them. They were gone to the south and to the gulf, and in the wake of their passing remained a soft echo of their vitality and purpose. The wind came up suddenly, and masses of leaves twisted, dipped, soared, and gathered in fading rushes, and piled in long lines, heaps, and circles. They were caught in bushes and whispered through the marsh at night; they were wet by rain and pressed to the earth. The varied hare crouched in his hide in a thicket and listened for the thump of a hare's foot in the rustling darkness which would tell him that other hares had begun a mad nocturnal game.

II

As the great forest was stripped of its leaf cover, it opened its interior to the pale sun, and a year of history was exposed with pristine clarity. Though the forest was constantly regenerating and resisting the inroads of fire from lightning and damage by snow, ice, hail, rain, and wind, the destruction appeared immense with the leaves gone.

It was an old forest, and scores of fallen limbs broke the symmetry of rising trunks and splayed branches. A dozen trees had fallen and created small clearings in which had sprung up dense growths of bracken, berry plants, fireweeds, and raspberries, and hundreds of young birches, balsams, aspens, and spruces. The forest had withstood the work of wood borers, the appetite of caterpillars, the hammering of woodpeckers, and the slow and deadly eating-away by funguses, whose life cycles, begun invisibly in open wounds, developed decay and death. The trees had survived the bud-hunting squirrels, the shoot-nibbling grouse, and the incessant appetite for bark of hares and deer.

The emerging skeleton of the forest showed that trees, like animals and insects, did not die of old age. Long before they reached the possible limits of life, they succumbed to winds or disease, thus making room for younger, healthier, and faster-growing trees. The body of a huge elm, which had fallen a dozen seasons before with an explosion that had roused the forest, came into sight north of the pond. Its fallen bulk had been afflicted immediately with fungus spores and mushrooms. Beetles burrowed into its trunk; its bark dried and fell off; and the forest animals nuzzled and clawed at the decaying wood, hunting the beetles and other insects. Moss grew prolifically, and gradually the tree fibers collapsed and spread out. The disintegrating hulk now lay among some red oaks that had gone almost purplish with the decay of their remaining leaves. The forest, cleansed of its old and sick, preserved an ordered and stately aspect, which was a perfect expression of life amid continuous death.

A great harvest of acorns, beechnuts, and butternuts was spread among the trees, and the squirrels were quickly gathering and hiding them. The old muskrat was dragging twigs and pieces of wood into the shallows near the swamp and piling them into a massive new winter nest near the dissolution of the cattails. Many forest animals were now fat and almost unwieldly, and this fat would sustain their bodies like food while they slept and would protect them against cold. The raccoons, on night excursions past the pond, stuffed themselves with anything edible. The skunks probed among the leaves and disturbed deer mice who were gathering stores of seeds, berries, and beechnuts. A mole dug with frenetic speed through earth that yielded him a constant harvest of food for his winter sleep. He could not, of course, see a late-migrating sharp-shinned hawk who had paused in a tree above to watch the spoor of disturbed soil created by his digging. The hawk was as hungry as the mole, and he dropped to the trembling patch of earth that concealed the subterranean creature and dug his talons deep into the soil. But either he miscalculated or the mole dug deeper into the soil, because after probing with his talons for a moment, he flew, victimless, back to a tree. Then he flew south. After a long wait, the mole resumed his rapid digging.

III

Ducks had been streaming into the pond and marsh, arriving singly, in groups, and in thundering thousands during the time of the migration. Their falling flights were variegated lines against bright-colored trees, and the sough and clatter of their wings echoed back and forth between the trees. They dropped into the water and spread out, ravenously searching for duckweeds, roots, and nymphs. Their arrival was almost continuous on some days, as feeding ducks called others down from the sky.

On gusty days, when the leaves were flipping across the sky in droves, the ducks banked against the marsh wind and made their approaches over the western end

of the pond, precise and graceful, coming in very fast and rocking in a gust of wind, then suddenly beating back, dropping their legs, and hissing into the water in one smooth motion.

On calm days, they arrived almost silently; only the acutest ear could hear the faint wheep of wings in twilight as they made low approaches, looking like spectral shapes that momentarily threatened to disappear and finally did so in the chuckling dark of the water.

On nights when the moon stood silently above the pond, sinuous lines of ducks passed across it, the fliers squawking and quacking as they moved ceaselessly to the south. Lone ducks called inquiringly and urgently from the marsh and the pond; these had been separated from their flocks or were solitary survivors of small family groups wiped out or scattered during the migration. The multicolored ducks flew through the vivid leaves, creating a bizarre mixture of reds and golds of sumac leaves; greens, blues, and browns of mallard ducks; reds, oranges, yellows, and greens of sugar-maple leaves; and reds and purples of canvasbacks and pintails. It was as much a mixture of types as colors, with shovelers, green-winged teals, harlequins, and eiders flying among the pale yellow birches and the fiery scarlet oaks. It was also a mixture of sounds: mutters, chuckles, clucks, squawks, purrs, and gulps. An occasional squawling scream came from the marsh.

The ducks overwhelmed the pond with numbers and created chaos throughout it as they shot down long necks, rooted deep in the mud, and ripped out roots and sleeping beetles and nymphs. They stirred up roiling clouds of silt and sometimes, when alarms were given, rose in a thundering body from waves and foam. The night brought no surcease from the confusion: clots of mallards took off in the moonlight and pintails came in to land, and wild calls sounded in Stygian gloom above.

Lost in the mass of numbers were individual ducks, such as the lone goldeneye who landed on the pond late one night and, avoiding other ducks, swam to the swamp and slept in shallow water there. He took off soon after dawn and disappeared. A golden scaup landed clumsily one afternoon and tipped forward into the water, nearly turning end over end. On recovering, she surveyed the pond with peculiar flips of her head as she tried to see all around her with one eye. A large patch of feathers had been torn from her back, exposing the purple flesh. She was a sick duck and sought refuge in the swamp.

The pond and the marsh enclosed, in these legions of ducks, a natural history of the north. It was a history of wings beating against blazing arctic sunrises and opalescent skies of evening. It was a history of bitter encounters with prowling arctic owls and of anguished alarms at midnight when the sun would not set and give the birds the security of night. It was a history of slashing attacks by kestrels and gyrfalcons and the searching flights of fish hawks and buzzards. It included the quick passages of curlews, ptarmigan, redshanks, and arctic foxes, and clots of tundra flowers and gray endless horizons.

The stillness of one early morning held the pond and its trees in utter quiet. A thick mist grew over the swamp and rose up its banks, and the albescent varied hare,

229

showing new fur growing on the healed scars of the
hawk's attack, looked down from a mound on the
opaque whiteness. He did not see a mink flanking him,
perhaps because the animal's outline was blurred in mist,
but suddenly the lean, umber creature was almost at
his nose. With a strain of muscle, he hurled himself
blindly from the mound, and his speed was so prodigious that he kept instinctively to smooth ground. This
led him to the mist, and he shot into it with the wind
hissing in his ears. Suddenly he hit water, and his momentum smashed him across it and ricocheted him into
a clump of rushes. He was erect in an instant, dripping
and sniffing. The mist was thick around him. Where was
the mink? The total silence emphasized the rapid rasping of air in his throat. He could smell the mink but
heard nothing except blood pounding in his head. He
was undecided and waited in the fog.

Beyond the mist, a triangle of birds flew silently
against a white cloud, and a crow humped along a horizon. The stillness persisted till after sunrise, when a pair
of white shapes came over the trees and heaved down
to land at the pond. The mute swans had come to rest
and feed, and by midday they were ready to leave.
They swam to the eastern end of the pond to take off
and together lunged forward, rising half out of the water, with wings beating and webbed feet stamping. But
they only rose slightly off the water and barely skimmed
the pond's peninsula before they had to swerve against
the trees and turn precipitately back to the water. For
birds of such bulk, a long take-off flight was essential,

230

and their miscalculation in landing on the pond was inexplicable. They tried taking off to the east, but again they failed. After three more abortive efforts, they floated in the calm water while the stillness settled over the pond again. In the evening, after waiting vainly for a wind that would lift them from the pond, they clumsily bumped down the exit stream leading to the marsh and, with its great expanses before them, labored into the air and headed south.

A fresh wind rose from the east in the late afternoon of the following day. Its gusty presence killed the remaining light, and rain started falling from its fleeting body. Thick gray clouds raced westward in darkness. The trees of the forest creaked and roared with a common voice. The migration went on, mysteriously audible through sudden barking calls from the swirling black air or visible as phantom shapes above the noisy trees. Small white-topped waves in the marsh ran among lines of ducks, now becoming serried under the stimulus of the wind; and some of the ducks took off and disappeared into the murk.

That night saw a sudden slackening of the migration of the multitudes. In the morning, lonely cries came across the marsh. Some birds even flew north into a growingly cold wind. The great honking cry of a goose pierced the air as the huge bird turned low over the pond and headed toward the marsh. Then he was over the pond again, calling harshly, and his cry sounded far in every direction but brought no reply. He flew west, his call sounding distantly once more, and then he was gone across the dying lines of flight.

the ice forms

In the final cycle of the year, ineffable melancholy, created by the contrast between life and lifelessness, spread over the pond. Even the light seemed drained of any vitality as it gloomily lit dead leaves whose fall colors had been leached from them and shone on dead gray goldenrods, gray formless swampland, the gray cattail leaves, and angular black trees. Dead nests straggled in their branches as silent manifestants of a surge of life now gone forever. A nuthatch droned out a call. The pond had burst open like an expansive blossom and had seeded and died and was now settling and rotting back into itself. The empty sky—the strangely empty sky, without the red-tailed hawk—promised only cold rain or snow.

The flat gray hush of the last season veiled everything in sleep so that a solitary raccoon, who kept returning to the pond, was an incongruous contrast. He should have been asleep with his fellows and with other sleepers like chipmunks, ground squirrels, and skunks. Perhaps, in his first experience of this season, he was kept wakeful by pervasive recollections of the year now gone. Perhaps he recalled the upwelling pond life that had taught him to hunt, the excitement of finding larvae under stones or catching frogs jumping desperately and vainly for the water, and certainly he remembered the wasp in the grass and the pain of the stings. Perhaps he sought the life and warmth that had disappeared and was baffled by the strangeness of this new season.

He prowled along the stream near the jutting peninsula and reached down among stones. A chill wind cut through a yellow sun that showed briefly, low on the horizon. He pulled a dragonfly larva from the water, stripped off some of its chitin and legs, ate the creature, and fastidiously washed his hands. He was feeling sluggish now, but for the moment he would resist his slowing metabolism. He looked dully at the diminishing yellow sun, and his breath steamed.

It was now penetratingly cold. A north wind brought

chill straight from the Arctic, and with its arrival the pond began its final change. The cold entered the ground and the tree trunks and slowed the last flickering movements of life in cell and fiber. It changed the appearance of the air and showed Polaris brilliantly clear in the night sky. Many stars were starkly white now, shining instead of twinkling, and they sparkled in the cold water. The moon was whiter. Its form had become small and remote, and gone was its encompassing influence of the past season, when it had revealed high-flying birds and zigzagging bats and the formless flying of legions of moths. Now it seemed to be receding, to be leaving the pond in the wake of the aerial migrants.

Familiar sights and sounds were gone and had been imperceptibly replaced. The purple night skies now contained constellations of unmatched form and beauty. Through bare branches shone the weak clustered lights of Coma Berenices. Deneb, in the constellation of Cygnus, shone brilliantly through gaps in the trees to the north. Low on the southern horizon lay Malus, Columba, and Fornax, which were dominated by the great stars of Orion, Betelgeuse, and Rigel, and the very lofty and majestic orange light of Aldebaran in the constellation of Taurus.

II

The temperature fell steadily and made the night dew viscous. Some great physical change was imminent. The pond terminated the northern declivity, invited cold air, and on this night heavy air drained into it from plateau and ridge, seeping down through the trees as invisible but palpable currents and eddies. It collected damply in hollows, and the following air glided over it and down to the pond. The snowshoe hare felt its bitter cold against his eyes as, in the dark, he sniffed into it for information, and under his feet, it sent thousands of creatures moving deeper into leaf cover and soil. It roused some shallow sleeping ants, and they clawed torpidly deeper into the mounds.

Shortly before dawn, the temperature dropped sharply below freezing, and the water in the flowing air froze wherever it touched. The dew froze, and a soft crackling rippled across the surface of the pond as water turned to ice crystals.

These were hexagonal and brilliant and floated uneasily in the water, tending to be attracted to one another. They stiffened the surface of the water and weaved toward each other, joined, then moved erratically to other groups of united crystals. Soon the pond reflected a veneer of thin ice that had a crystalline structure interlaced with millions of lines moving jaggedly in every direction.

The ice materialized out of the air itself. It froze the cattail stalks lightly, and they rattled softly when stirred in the wind. It spread round the edge of the pond and melted and re-formed and moved toward the center. In one night, it covered the pond and thickened to catch and freeze many creatures. It brought death to some and protection to others. Five hundred freshwater snails were enclosed in ice among the reed beds

235

and they slept unharmed. A dozen back swimmers were frozen to a bur-reed stalk, and they would die. Ten song sparrows slept in the old cattails, and their feet melted the ice forming around them. A woodpecker slept in a hole he had dug in rotten wood in a willow trunk; icicles hung thickly over the entrance.

III

The change of season momentarily immobilized the pond till northern travelers began arriving. They were migrants like those who had preceded them at the pond, but their presence would be more transitory because they would have to cover vast territories to sustain themselves. A steady shower of pine-cone petals fell out of a pine tree to the lightly iced pond. A flock of handsome crossbills were tearing out the seeds with powerful, wood-cutting crossed beaks. The pond's heavy cone crop attracted other birds who were refugees from the failure of many trees in the north to produce a good crop of seeds. Cone-feeding pine siskins appeared in the evergreens, along with vagrant flocks of redpolls, whose kind spanned the roof of the world. Sturdy grosbeaks arrived and joined the crossbills in feeding. Flocks of goldfinches came out of hushed skies in clouds of evanescent prettiness, undulating over the ice and studding bushes like greenish dewdrops. The pond emerged once again as the great provider, a fountainhead that even attracted twittering flocks of snow buntings who had molted into pure white plumage in the Arctic and then had fled the food shortage there.

236

During this transitional period that overlapped the seasons, many creatures would be eliminated from the pond. A shrike had chosen the pond as the center of his winter range, and he looked harmless enough, robin-sized and hunched up grayly on an elm twig. He watched a white-breasted nuthatch circling the elm trunk and then dropped down and killed him with a lack of haste that concealed his perfectly developed judgment of reactions and attention.

The cold slowed perceptions and sealed the country against the odd vagrant birds skulking in the marsh. Most of them flew alone and silently now, in the realization of their plight. They were victims of the hunt or of diseases that had stopped their instinctive flight south. Now the instinct had gone, and they faced the insuperable difficulties of a season none of them understood.

A Canada goose swam alone in the marsh, with the fearful awareness of an injured, permanently crippled wing, which gave her a lopsided, staggering aerial gait and imperiled her constantly. She flew infrequently, but some alarm in the marsh sent her pumping awkwardly into the air, and she appeared suddenly over the pond's southern elms, for a moment magnificent and powerful, then slipping in distress down to the thin ice. She smashed through it and bobbed up, erectly scanning the enclosing trees with an urgency born of premonition. A fox looked at her through some dry reeds, and the goose broke a jagged path through to the center of the pond. Her fear rose suddenly and overcame her instinct to regain strength, and with a crash of wings she

237

plunged forward to the pond's peninsula. Underwater, the crackling thud of wings sent many wakeful aquatic creatures into hiding. With more than half the length of the pond gone, she was almost into the air, though still stamping feet on the ice. She swung widely over the peninsula but now was rising only under the impetus of her desperation, and it was obvious that she would never clear the tree-top line. She disappeared into the forest beyond the swamp, and the fox remained motionless in the reeds, listening intently to wings hitting branches and a muffled impact and uncontrolled flailing. He twitched his whiskers and trotted along the bank of the pond to the noise.

Other vagrants came to the pond, sometimes in the middle of chill black nights. A mallard landed and took off one night, using the unfrozen lines of broken ice left by the goose. A fish-eating duck, a merganser, dived under the ice and swam deeply through the pond to catch a bass and a bullhead.

IV

As the pond was tightly clamped into winter, the last faint undertones of previous seasons died away completely. The ice thickened, and some ducks and coots swooped down to it, then veered away, realizing that it was too thick to break. Only in the marsh was there a small black gash of unfrozen water, where a deep spring welled up.

As the ice spread and the first real snow drifted down, even this small stretch of water was threatened by ice.

One night it formed around a sleeping coot who was enervated by arduous hunting, and growing crystals stuck to his primary wing feathers. But he slept on in an aftermath of despair.

He awoke shortly before dawn and was instantly aroused; he was unable to move. He jerked his wings, got one free from the ice, and kicked his legs, but he could not move his trapped body. He stretched his neck

and looked to the graying east. He dreaded the daylight, which was synonymous with marsh minks, and flapped his free wing futilely against the new ice. The other birds were awake and alarmed by the sound and looked into the gloom for a response to it. A mallard took off and circled inland. As the gray tint of dawn lightened behind the angular trees around the pond, a tiny saw-whet owl twisted his head backward to the sound of a strangled screech from the marsh. A moment later, ducks and coots fled over the pond; it was soon dawn, and another winter day was beginning.

A wind came over the deepening ice, so cold it drove most living things into hiding. This was perhaps the actual moment at which the year ended. The minks retreated into shelters in patches of drifted snow or into disused or commandeered muskrat shelters. The old muskrat, dozing alone in his nest, suddenly twitched his grizzled snout and died. The cold drove the snowshoe hare into a refuge in a hollow tree half buried in snow, and he looked across the pond, shivering. It sent bobwhites and grouse into huddled refuge in thickets and behind rocks. It froze sparrows to death that night and closed the remaining gash of open water. It drove the last of the vagrant birds into oblivion.

By the morning of the last day, the pond was graven in the final shape of this absolute season. The wind died, and light snow fell from a lowering sky. Against this was fixed the eternally wheeling bird. But it was not the red-tailed hawk. It was a raven. He had survived with equanimity the changing seasons and the death of his solitary rearling, killed by a weasel, and the disappearance of his mate. He had come to the pond and marsh as was his custom and looked down on them for the thirteenth season in his life.

He had watched the withdrawal of life from the north and had flown with some of the southward-streaming ducks. He had seen salamanders swimming up streams and toads jumping through woodlands and a profusion

of life pouring past him to the south. He had impassive-
ly watched the arrival of the northern buntings and
finches and a pair of arctic owls with whom he would
have to share some of the hunting at the pond.

The raven was one of the great watchers. His long experience was preserved in a fund of memories that made
him wilier and more watchful the older he grew. Now
that the year of the pond was fading into the past, he
personified that year and others before it. His memory
was sharp, but he was not unique in having this faculty.
That pond had it, and that was the ultimate secret. A
haunting echo of the past was preserved within countless quiescent cells, within protoplasm and egg, within
brain and bone. It was preserved in pupating grubs
and sleeping moths and in the fertilized unlaid eggs of
wasps and bees. Each egg, however small, carried mes-
sages from the previous year and from years, centuries,
millennia, and eons before that.

In this memory was every detail of the spring awakening: the reaching for space, light, expression; the languorous heat of summer days; the slow waning metabolism of the last season; and the long sleep away from
the sun. In every speck of living matter, there was this
memory of the indestructible life force of earth.

The raven turned into northern haze and disappeared;
the sleepers slept on; and the pond moved to infinity.

A NOTE ABOUT THE PRODUCTION OF THIS BOOK

The typeface for the text of this special edition of *Watchers at the Pond* is Century Expanded. It was photocomposed at Time Inc. under the direction of Albert J. Dunn and Arthur J. Dunn.

x

PRODUCTION STAFF FOR TIME INCORPORATED: John L. Hallenbeck (Vice President and Director of Production), Robert E. Foy, Caroline Ferri and Robert E. Fraser.